Introduction to Semiconductor Device Yield Modeling

The Artech House Materials Science Library

Introduction to Semiconductor Device Yield Modeling

Albert V. Ferris-Prabhu
IBM Corporation

Artech House
Boston • London

Library of Congress Cataloging-in-Publication Data

Introduction to semiconductor device yield modeling / Albert
V. Ferris-Prabhu.
   p.   cm.
   ISBN 0-89006-450-4
   1. Semiconductors—Defects. 2. Semiconductor industry—Quality
control. I. Title.
TK7871.85.F46 1992   92-10875
621.381'52—dc20   CIP

© 1992 ARTECH HOUSE, INC.
685 Canton Street
Norwood, MA 02062

International Standard Book Number: 0-89006-450-4
Library of Congress Catalog Card Number: 92-10875

10 9 8 7 6 5 4 3 2 1

To the memory of my parents

Contents

Preface

The purpose of this book is to provide an introduction to yield prediction for new practitioners of the art. Though based solidly on probability theory and required to meet the inexorable demands of economic viability, yield prediction still has areas of uncertainty where it is often necessary to use informed experience to interpret experimental data and to fathom business needs. Nevertheless, it was felt that the time is appropriate to present an introductory overview of an increasingly important subject.

In this spirit, Chapter 1 introduces the concept of semiconductor yield, discusses the need for accurate yield predictions, and comments on the fabrication of semiconductor devices and the defects that lower yield. Chapter 2 describes how to calculate the probability that a potentially fatal defect will indeed be so, followed in Chapter 3 by a discussion of the effect on this probability of different assumptions about the distribution of defect sizes. Chapter 4 presents expressions for counting the number of ways in which events can be distributed over all possible states when their occurrence is governed by different distribution laws. This leads into Chapter 5, which defines yield as the probability of obtaining defect-free chips and presents several equations that do so under different assumptions about how the defect density varies across and between wafers. Chapter 6 discusses the idiosyncrasies of inferring the defect density from yield data, presents rules for scaling these densities from an existing product to the values expected on a proposed product, and revisits the often maligned Poisson yield equation. In Chapter 7 the preceding material is woven together to present a description of how to predict yield for a proposed new product, based on information about its design, as well as all available data from existing product. This is extended in Chapter 8, where expressions are presented for the yield of array product that have on-chip redundancy. Chapter 9 shows that it is possible for more than one yield equation to provide reasonably accurate yield predictions with the use of appropriate rules to scale the defect density, and suggests that in such cases the yield equation that is more computationally efficient be used. Chapter 10 concludes with a definition of productivity, which in the last analysis is what counts, and briefly discusses how the productivity of different proposed designs might be assessed.

Because this book is addressed to those working in the field, examples and problem sets have not been included. However, if there is interest, a second edition can be provided that does, so comments from readers would be appreciated. In such a fast-growing field, it is not possible to refer to every relevant publication, but I have tried to cite the more important references that I am aware of, including those that either provide a different point of view or confirm an existing one. Although much effort has been expended on trying to eliminate defects, it is improbable that this book will be completely free of them. If there are any, I hope they will not be fatal.

It is with great pleasure that I acknowledge the support and assistance of my wife Joy and son Arjun; without their love and encouragement it would have been impossible to commence, let alone complete, this book. I am deeply grateful to them.

Finally, I wish to make clear that I am solely responsible for the contents of this book. The book was not prepared by or for IBM, and does not necessarily represent the views of the IBM Corporation.

Chapter 1
Yield

1.1 INTRODUCTION

This chapter explains what yield means and why it is important to be able to predict it accurately. We define what is meant by cumulative yield and describe each of the terms used to evaluate it. We include brief comments on device fabrication and a short discussion on one method of categorizing defects that arise during fabrication. We also explain that the largest contributor to the cumulative yield loss is the loss of device yield, loss that occurs when potentially fatal defects generated during device fabrication become fatal to device operation, and we point out that a component of yield prediction is to determine the probability of a defect becoming fatal.

1.2 DEFINITION OF YIELD

Yield can be defined as the ratio of the number of items that are usable after completion of production to the number at start of production that have the potential to be usable. This definition is as applicable to natural products, such as corn, wheat, or soybeans as it is to manufactured products, such as automobiles, toasters, or semiconductor chips.

In the semiconductor industry, ever since the invention of the transistor, there have been two seemingly contradictory trends. In terms of constant dollars, the cost of manufacturing has increased steeply. At the same time, the cost per circuit has decreased even more steeply. The reason for this seeming anomaly is improved yield.

A typical semiconductor device consists of a rectangle of silicon with edges anywhere between 1 mm and about 15 mm. However, devices are not manufactured singly. They are manufactured in large numbers on a wafer, in batches of hundreds of wafers a day. A wafer is a circular slice of silicon, typically about 100 to 125 mm in diameter, though in some cases it may be as large as 200 mm in diameter or even larger. A sequence of operations is performed on the wafer, at the end of which the wafer consists of a number of identical rectangular regions called chips or dies, separated by a region called the kerf. Later, when dicing the wafer to obtain individual chips, which are packaged into modules for use, the saw blade cuts through the kerf. For this reason, the kerf is either left blank or used for special test structures used for testing prior to completion of fabrication and final test. The yield per wafer is the fraction of all potentially usable chips that is actually usable. As wafers get larger, manufacturing complexity, and therefore cost, increases. However, the number of chips per wafer also increases, and with advances in manufacturing techniques, more circuits can be fabricated in the same area. The more functional chips obtained (i.e., the higher the yield), the lower the cost per circuit will be. However, because of the expense involved in designing and manufacturing new and more powerful devices, it is essential for strategic as well as tactical reasons to be able to predict the expected yield of a proposed product. An example of a strategic need would be the need to assess the yield in, say, the year 2000 of a potential 2-inch-by-2-inch analog chip with both memory and logic features, four levels of metallization, and 0.1 μm minimum design features in a gallium arsenide (GaAs) technology, so that appropriate plans can begin to be put in place should the analyses suggest that such a product will be economically viable. A tactical need, on the other hand, would be to estimate the number of wafer starts needed next quarter to meet the committed volume of a 5-mm-by-5-mm memory chip with two levels of metallization and 2-μm ground rules in a standard silicon technology. In both cases, accurate yield predictions enable informed decisions to be made in a timely manner. Improved productivity with attendant cost reduction is central to a successful business. A simple example will illustrate these remarks.

Consider a manufacturer who has designed a 5-mm-by-5-mm chip for fabrication on a wafer 125 mm in diameter and is ready to go into production for a customer who is about to place an order for 5 million chips. At least 425 such chips can be fabricated on a wafer, and if each chip on a wafer is assumed to be functional, 10,000 wafers will be needed. If 100 wafers can be started daily, seven days a week, it will take at least one hundred days to start all the wafers needed, plus the manufacturing cycle time. Based on this estimate a price of twenty-five dollars per chip is quoted, on which basis each day's production of finished chips represents an inventory of over one million dollars.

But suppose that on average only 50% of the chips on a wafer are functional. Then 20,000 wafer starts will be needed to be run over 200 days, during which the line may well not be available for other products. If the manufacturer is unaware of such a possibility and has made no contingency plans to address such a situation should it arise, not only will deliveries be delayed, resulting in loss to the customer, but because the actual costs will be larger than the estimates on which the price is based, the manufacturer may undergo severe losses. Improved productivity and its attendant cost reduction being central to a successful business, it is clear that accurate yield prediction is necessary in developing a viable production program. Yield prediction is central in the determination of cost.

1.3 CUMULATIVE YIELD

Integrated circuit semiconductor device modules are the end result of three sets of manufacturing operations: growing of the wafer, fabrication of the devices on the wafer, and packaging into chip modules into which the wafer is diced. Many device manufacturers buy raw silicon wafers and process them so to produce a particular product, which may be a memory device, a logic device, or an application-specific device. The fraction of wafers begun that successfully have devices fabricated is called the wafer process yield. The fraction of devices on a fabricated wafer that pass final test is wafer test yield, or device yield. And the fraction of packaged devices that pass module final test is called module test yield. The cumulative yield is the product of wafer process yield, wafer test yield, and module yield. These three components are represented schematically in Figure 1.1.

The cumulative yield (Y_{cum}) is equal to the product of the wafer process yield (Y_{WP}), the wafer test yield (Y_{WT}), and the module test yield (Y_{MT}), each of which is defined below.

$$
\begin{aligned}
Y_{WP} \ &= \ \text{wafer process yield} = w_{out}/w_{in} \\
w_{in} \ &= \ \text{wafers started} \\
w_{out} \ &= \ \text{wafers completed} = w_{in} \times Y_{WP} \\
Y_{WT} \ &= \ \text{wafer test yield (or device yield)} = \text{gcpw}/\text{cpw} \\
\text{gcpw} \ &= \ \text{chips per wafer passing test} \\
\text{cpw} \ &= \ \text{potentially good chips per wafer} \\
c_{in} \ &= \ \text{chips in to test} = \text{cpw} \times w_{in} \times Y_{WP} \\
c_{out} \ &= \ \text{chips passing test} = c_{in} \times Y_{WT} \\
Y_{MT} \ &= \ \text{module test yield} = m_{out}/m_{in} \\
m_{in} \ &= \ \text{modules in to test} = c_{out}/\text{cpm} \\
\text{cpm} \ &= \ \text{chips per module} \\
m_{out} \ &= \ \text{modules passing test} = m_{in} \times Y_{MT} = (w_{in} \times \text{cpw}/\text{cpm}) \times Y_{cum} \\
Y_{cum} \ &= \ Y_{WP} \times Y_{WT} \times Y_{MT} \\
P \ &= \ \text{productivity} = \text{packaged chips out/wafer started} = \text{cpw} \times Y_{cum}
\end{aligned}
$$

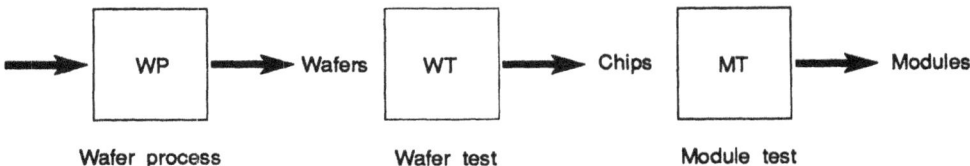

Figure 1.1 Components of the cumulative yield.

Productivity, which is the number of packaged chips available per wafer started, can be increased by using larger wafers (which have more potentially usable chips on them) and by increasing the cumulative yield. Of the three components of cumulative yield, module test yield is usually the highest, being very close to 100%. The wafer process yield can range from about 60% in the beginning stages of development to about 80% or even 90%. The most important component of the cumulative yield, and thus of the productivity, is the device yield, also called wafer test yield.

The essence of device yield prediction is to be able to obtain a numerical value for the probability $p(0,\lambda)$ of obtaining a chip with no fatal defects on it, when there are on average λ fatal defects per chip. The form of the yield equation that gives this probability depends on the spatial distribution of the fatal defects, and the average number of these per chip depends on the size distribution of all potentially fatal defects. Both these topics will be treated in more detail, as they play an important role in yield modeling. However, it is advantageous to summarize the essence of device fabrication because the defects that cause yield loss occur during one of the several fabrication steps.

1.4 DEVICE FABRICATION

As devices become smaller, the tools needed for their fabrication become more complicated and expensive, and the environment in which the wafers are processed has to be ultraclean. However, conceptually the process is quite straightforward. The blank wafer is covered with a material impervious to the penetration of selected impurities, which provide the desired electrical characteristics. Using a mask that protects selected areas, openings are made in the unprotected areas into which the desired impurities are deposited, and these areas are then covered with a suitably impervious material. Depending on the type of impurity and how it is deposited, the wafer may be subjected to a heat cycle to allow it to diffuse to the desired depth. This sequence of covering the wafer surface, opening selected areas, depositing the desired material, covering the opening, applying heat treatment if needed, and starting the procedure again is essentially repeated until precisely defined regions in successive layers of the wafer contain the material providing the electrical characteristics desired. To provide contact between the desired regions, one or more layers of metal, separated by an insulator, usually silicon dioxide, are deposited, and the entire surface is finally covered with an insulator on which metal or polysilicon is deposited in selected regions to serve as contacts for electrical continuity between the devices built into the wafer and the external world. Each of the dies on the wafer is tested, and the location of the chips that fail test are recorded. The wafer is diced, and the good chips picked and stored until they are ready to be packaged. This is schematically shown in Figure 1.2.

Raw wafer from stock
|
-------------------- Wafer processing
|
Epitaxy
|
Oxidation ◄-----------------┐
| |
Photoresist |
| |
Expose and develop photoresist   repeat as needed
| |
Etch openings where needed |
| |
Diffuse or ion implant impurities --------┘
|
Deposit metal layer
|
Remove unwanted metal
|
Probe test chips
|
Scribe, test, dice chips   **Wafer test**
|
Select the good chips
|
------------------------ **Packaging**
|
Mount chips in modules
|
Test assembled module   **Module test**
|
Send good modules to stock

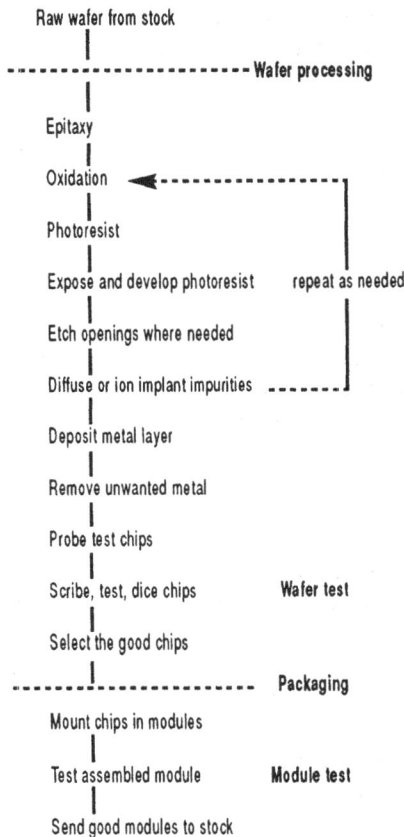

Figure 1.2 Schematic of device fabrication.

1.5 CLASSIFICATION OF DEFECTS

The sequence of processes that occur in fabricating devices on a wafer is grouped into operations conveniently categorized by each mask step. The mask steps are either labeled with letters (e.g., A, B, C, etc.), or by letters that reflect their purpose, such as, SC for the subcollector mask, R for the reach through, M1 for the first metal mask, V for the mask that allows vias to be etched in the insulator over the metal, and so on.

It is almost impossible to build a product with no defects, and there are various ways of categorizing the defects that occur during fabrication. The simplest is to categorize them as photo, leakage, or miscellaneous defects. Photo defects can be categorized in terms of missing or extra patterns. Missing patterns typically break a path that is meant to be conducting. Extra patterns (patterns that are inadvertently formed) connect two regions that were meant to be electrically separate. Leakage defects are those that result in leakage of charge from one region to another because of, for example, contaminant inclusions, dislocations in the silicon structure, or pinholes in the insulator between two regions meant to be electrically separate. Among miscellaneous defects may be grouped scratches, foreign material, field size errors, and so on.

A more detailed way is to categorize the defects that occur at each mask level as being due to extensions, scratches in the resist, holes etched in the silicon, residual oxide, floating resist, undersized or oversized via holes, floating or unlifted metal, dislocations in the silicon, junction leakage, pinholes in the dielectric, and so on. Although these are some of the basic types of defects that can occur, each technology has defects that are peculiar to it. These are usually discovered as the technology is being developed and work is done to determine their incidence and causes.

From the point of view of yield prediction, it is often convenient to categorize defects as being gross, parametric, or random defects, with the latter characterized as photo, leakage, or miscellaneous defects. Depending on the level of detail needed, these may in turn be categorized more specifically.

Gross defects are those that are relatively large with respect to the size of a circuit or a chip. Examples of these are scratches, damage due to handling or placement of probes or other tools, or large strips of metal or photoresist that have not been completely removed. Gross defects almost always result in yield loss.

Parametric defects are those that affect the device electrical parameters. For example, an uneven concentration of dopant across a chip can result in differing values of the resistance of two resistors meant to be identical. This type of defect does not result in any physically observable damage to the chip, though it can result in the chip not functioning as it should. Parametric defects can, and sometimes do result in yield loss, but more frequently they lead to reliability problems when after a period of time the device fails to function as designed.

Random defects are defects that have a *chance* of occurring. A numerical measure of this chance is called the *probability*. However, the manner in which defects are distributed over the surface of the wafer is not always clear, though there is much evidence to suggest that there is a greater likelihood of defects occurring near the periphery of the wafer than in regions near the center. Random defects that result in yield loss are referred to as fatal defects, or faults, and the likelihood of a defect becoming fatal is called the fault probability. The next chapter shows how to calculate the fault probability.

1.6 REFERENCES

[1]   R.A. Colclaser, *Microelectronics: Processing and Device Design*, John Wiley, 1980.
[2]   S.K. Gandhi, *VLSI Fabrication Principles*, John Wiley, 1983.
[3]   S.M. Sze, ed., *VLSI Technology*, McGraw-Hill Series in Electrical Engineering, McGraw-Hill, 1983.
[4]   I. Brodie, and T.J. Muray, *The Physics of Microfabrication*, Plenum Press, New York and London, 1982.

Chapter 2
Fault Probability

2.1 INTRODUCTION

This chapter explains the distinction between defects, which are potentially fatal, and faults, which are defects that are actually fatal. It defines the fault probability kernel, average fault probability, and defect-sensitive area (or critical area); shows a method of obtaining the kernel for simple patterns; and derives formal expressions for the average fault probability for simple patterns.

2.2 DEFECTS AND FAULTS

Random defects may be *defined* as any deviations from design that have a chance of occurring. The presence of a pattern where it should not be is a defect, as is the absence of a pattern from where it should be. To *determine* when something actually *is* a defect is not as straightforward because, at least as far as yield prediction is concerned, a defect is of no consequence unless it results in yield loss, and defects may sometimes not be distinguishable as defects until they result in yield loss. Defects that result in yield loss are called fatal defects, or faults. The numerical measure of the likelihood of a defect being fatal (i.e., becoming a fault) is called the fault probability, and the area within which the occurrence of a defect makes it a fault is called the defect-sensitive area, or critical area [1]. The fault probability [2] is the ratio of the defect-sensitive area, in which the occurrence of a defect *will* be fatal, to the total area in which the defect *can* occur.

Random defects that can become faults may be categorized as being of two kinds: point defects and defects that have finite size, here referred to as spatial defects. Point defects include pinholes in insulating layers, inclusions of conducting contaminants of atomic size, or dislocations in the crystal structure, any one of which may be fatal, depending on where it occurs. Spatial defects can result in missing or extra patterns. A missing portion of a conducting pattern will be a fault if it results in an open circuit. An extra pattern will cause a short-circuit fault if it bridges two patterns meant to be electrically separate. However, for spatial defects to become faults, not only must they occur where they *can* be fatal, but they must also be of at least a minimum size, as we shall see.

2.3 FAULT PROBABILITY FOR POINT DEFECTS

If a point defect occurs where it *can* be fatal, it *will* be fatal. For example, a pinhole occurring in the insulator separating two levels of metal will necessarily be fatal if it occurs in the overlap area shown in Figure 2.1.

The fault probability Φ of such a defect is the ratio of the defect-sensitive area A where it will be fatal to the total area A_T of the chip; that is,

$$\Phi = \frac{A}{A_T} \tag{2.1}$$

The exact magnitude of the defect-sensitive area for this type of defect is computed from the chip layout of all regions for which there is an insulating plane between two conducting planes. Such computations are tedious.

If N_D is the total number of such random defects within the area A_T, the number of them that become faults is given by

$$\lambda = N_D \Phi \tag{2.2a}$$

which can also be written as

$$\lambda = AD \tag{2.2b}$$

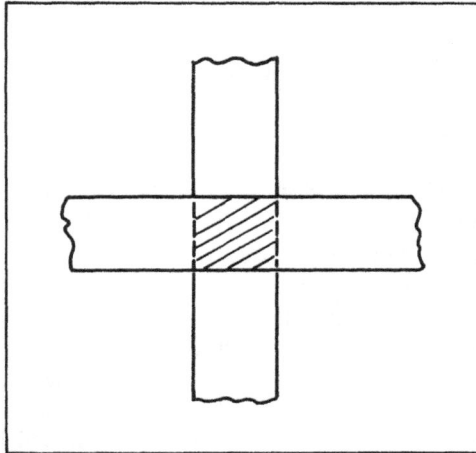

Figure 2.1 Defect-sensitive area for the insulating plane between two metal levels.

where the defect density D is the ratio of all defects of this kind to the total area in which they lie; that is,

$$D = \frac{N_D}{A_T} \tag{2.3}$$

2.4 SINGLE-PATTERN FAULT PROBABILITY

The probability of a spatial defect being fatal depends on its location and its size. To obtain expressions for the fault probability of such defects, it is helpful to start with an idealized case [3–5]. Figure 2.2 shows a pattern consisting of a single metal line of width w, much less than its length L, deposited on an insulating substrate of width X and length Y.

A defect in the photo mask, or a piece of photoresist that has not been completely removed prior to deposition, or the inclusion of some foreign material during

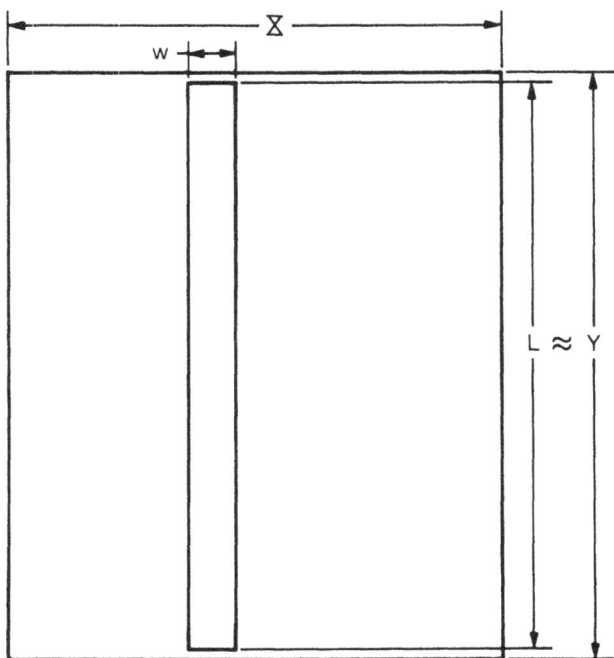

Figure 2.2 A single pattern. The pattern of length L and width w lies in a region of length Y and width X.

deposition, can result in the pattern being partly or completely broken, depending on the size of the defect. If the pattern is completely broken or narrowed to less than some minimum width, it will not conduct current, and the resultant open circuit will be a fatal defect, or fault, that results in yield loss. Assume that an open circuit results only if the pattern is completely broken, and for convenience consider such defects to be circular. For such a potentially fatal defect to become a fault, its diameter x must be at least as large as the width w of the pattern in order to break it completely. Furthermore, defects of size $x \geq w + X$ will always be fatal, no matter where they fall. But as shown in Figure 2.3, a defect of diameter larger than the

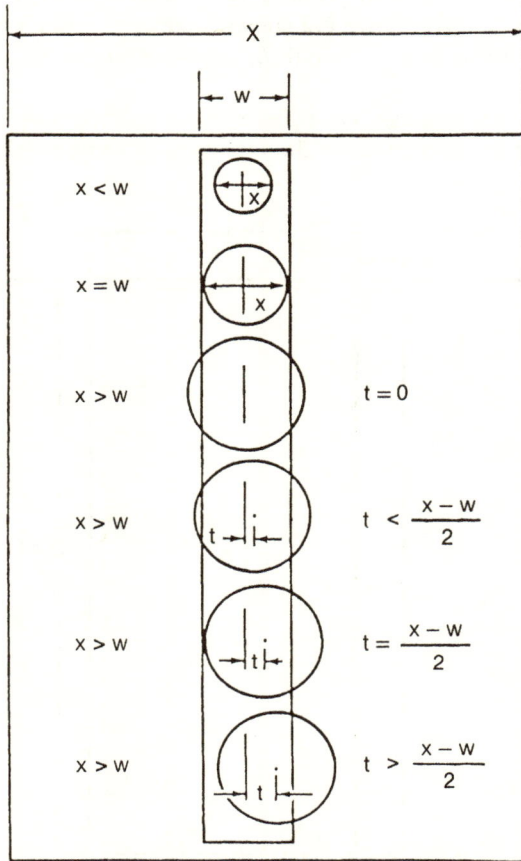

Figure 2.3 Critical interval for defects of size $w \leq x \leq w + X$. The defect will be fatal only if its center falls in an interval $t \leq x - w$ from the centerline of the pattern.

pattern width w but less than $w + X$, where X is the width of the region in which the pattern lies, can break the pattern completely only if the distance t of its center from the centerline of the pattern does not exceed half the difference between its diameter x and the pattern width w; that is, the center must fall in an interval of width $(X - w)$ orthogonal to the centerline of the pattern.

The ratio of the critical interval within which such a defect *must* fall in order to be fatal to the width X of the substrate on which it *can* fall is called its fault probability kernel $K(x - w)$ and is shown in Figure 2.4.

$$K(x - w) = 0, \quad x \le w \tag{2.4a}$$

$$K(x - w) = (x - w)/X, \quad w \le x \le w + X \tag{2.4b}$$

$$K(x - w) = 1, \quad w + X \le x \tag{2.4c}$$

The fault probability $K(x - w)$ for a single defect of size x on a pattern of width w is called a *kernel*, since it is structurally similar to the term called a kernel in the theory of integral equations. In fact, Equation (2.5), in which $K(x - w)$ appears, is a Fredholm homogeneous linear integral equation of the first kind, in which the average fault probability Φ, or the defect-sensitive area to which it is related, is

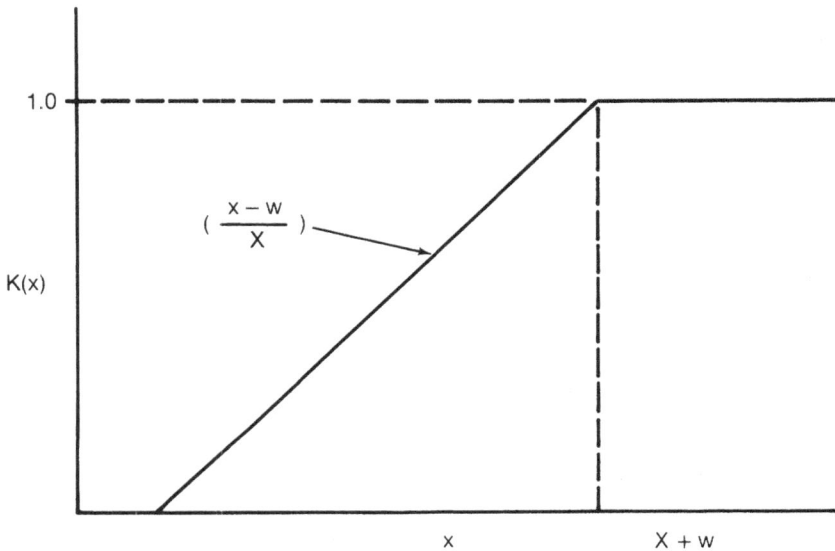

Figure 2.4 The open-circuit fault probability for a single pattern.

empirically known, the size probability density function $S(x)$ needs to be determined, and $K(x - w)$ serves as the kernel [6, 7].

The average fault probability of all defects that can cause this type of fault is the product of the fault probability kernel $K(x, w)$ and the fractional number of defects of size x integrated over defects of all sizes. If the defect size probability density function is $S(x)$, then the average fault probability is given by

$$\Phi = \int_0^{x_M} dx S(x) K(x - w) \tag{2.5}$$

where x_M is the largest size defect that can occur. In practice, the lower limit on the integral is w, because defects smaller than the pattern width are not fatal.

The area sensitive to defects that can be fatal is approximately $(x - w)Y$, or

$$A = K(x - w)XY = K(x - w)A_T \tag{2.6}$$

where A_T is the total area of the substrate on which this pattern lies.

The total defect-sensitive area for this pattern of width w is

$$A(w) = A_T \int_0^{x_M} dx S(x) K(x - w) \tag{2.7}$$

If N_D is the number of defects of this type in the area A_T, the average number of them fatal to this pattern is

$$\lambda = N_D \Phi = AD \tag{2.8}$$

where the density of defects of this type is $D = N_D/A_T$.

2.5 TWO-PATTERN FAULT PROBABILITY

For a region with more than one pattern, there are two types of fatal defects. Those that break a pattern, resulting in an open-circuit fault, and those that connect two patterns meant to be separate, resulting in a short-circuit fault.

2.5.1 Short-Circuit Fault Probability

Consider the two patterns shown in Figure 2.5, each of length $L \simeq Y$ and width $w \leq X$, with separation s, lying on an insulating substrate of dimensions X, Y. For

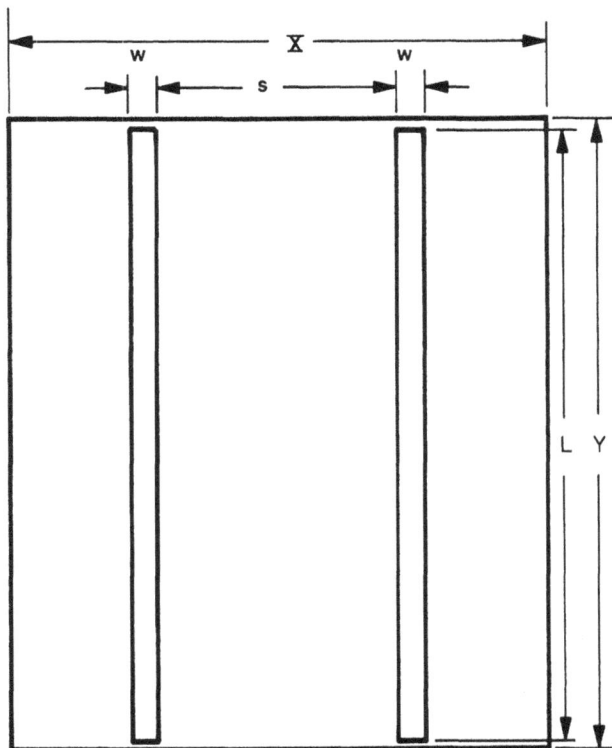

Figure 2.5 Two patterns, each of length L and width w, with separation s.

later convenience, let us define a second separation s' as the separation between the outer edges of the patterns or, equivalently, the separation between a pattern on one chip and the corresponding pattern on an adjacent chip; that is,

$$s' = X - (s + 2w) \tag{2.9}$$

The short-circuit fault probability kernel for this case is obtained in direct analogy to that of the single-pattern case, with the space s between patterns replacing the width w of the pattern; that is,

$$K_s(x - s) = 0, \quad x \le s \tag{2.10a}$$

$$K_s(x - s) = (x - s)/X, \quad s \le x \le s + W \tag{2.10b}$$

$$K_s(x - s) = 1, \quad w + X \leq s \tag{2.10c}$$

The average short-circuit fault probability is

$$\Phi_s = \int_0^{x_M} dx K_s(x - s) S_s(x) \tag{2.11}$$

where S_s is the density function for short-circuit inducing defects.

The short-circuit defect sensitive area is

$$A_s = A_T \Phi_s \tag{2.12}$$

and the average number of short-circuit faults is

$$\lambda_s = N_{D,s} \Phi_s = A_s D_s \tag{2.13}$$

where $N_{D,s}$ and D_s are the number and density, respectively, of defects that can cause short circuits.

2.5.2 Open-Circuit Fault Probability

The open-circuit fault probability kernel can be obtained by considering that there is an adjacent chip with separation s' between patterns on adjacent chips.

We note as before that defects smaller than the pattern width w are not fatal. For defects larger than w but less than $s + 2w$, the critical interval for both patterns together is the sum of the critical interval for each pattern. But, as shown in Figure 2.6, defects larger than $(s + 2w)$ can open both patterns at the same time; that is, the critical interval for each pattern overlaps that of the other by the interval

$$x_{ov} = x - (s + 2w), \quad w \leq x \leq s + 2w \tag{2.14}$$

giving the net critical interval

$$x_c = 2(x - w) - x_{ov} = x + s \tag{2.15}$$

Defects larger than $(s' + 2w) = X - s$ are fatal no matter where they fall. Thus, the open-circuit fault probability kernel for these two patterns is

$$K_o(x - w) = 0, \quad x \leq w \tag{2.16a}$$

$$K_o(x - w) = 2(x - w)/X, \quad w \leq x \leq s + 2w \tag{2.16b}$$

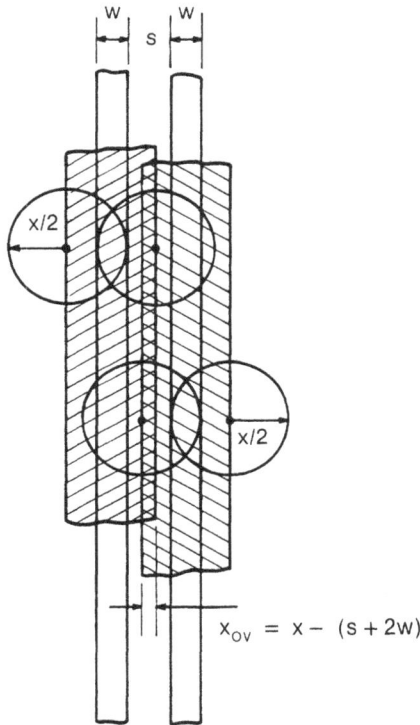

Figure 2.6 The critical intervals meet for defects of size $x = s + 2w$. For larger defects, the critical intervals overlap by the extent $x - (s + 2w)$.

$$K_o(x - w) = (x + s)/X, \quad s + 2w \leq x \leq X - s \qquad (2.16c)$$

$$K_o(x - w) = 1, \quad X - s \leq x \qquad (2.16d)$$

As indicated by Equations (2.16b) and (2.16c) and shown in Figure 2.7, after the defect reaches the size $(s + 2w)$, the gradient of the fault probability kernel drops to half the value it had for defects smaller than that size. It then stays constant till it reaches unity for defects of size $X - s$. The average two-pattern open-circuit fault probability is

$$\Phi_o = \int_0^{x_M} dx K_o(x - w) S_o(x) \qquad (2.17)$$

where S_o is the density function for open-circuit inducing defects.

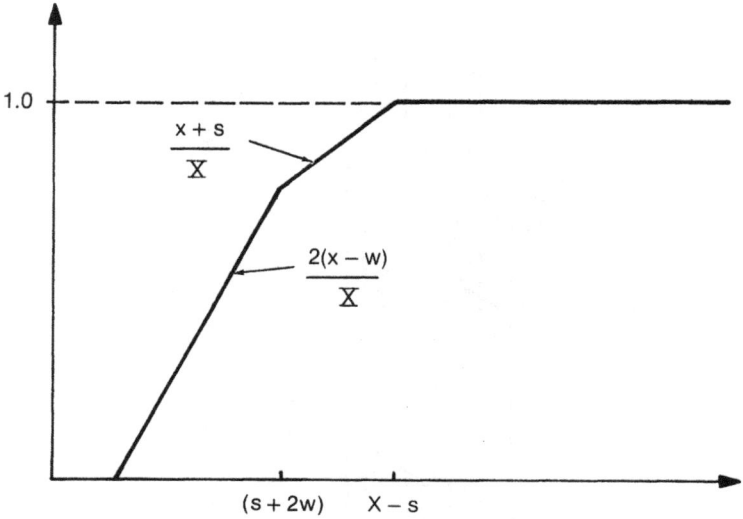

Figure 2.7 Two-pattern open-circuit fault probability. The gradient of the fault probability kernel decreases when the defect is large enough to cover both patterns.

The open-circuit defect sensitive area is

$$A_o = A_T \Phi_o \tag{2.18}$$

and the average number of open-circuit faults is

$$\lambda_o = N_{D,o} \Phi_o = A_o D_o \tag{2.19}$$

where $N_{D,o}$ and D_o are the total number and density of defects, respectively, that can cause open circuits.

For this two-pattern case, the average number of fatal defects of both kinds is

$$\lambda = \lambda_s + \lambda_o \tag{2.20}$$

2.6 MULTIPLE-PATTERN FAULT PROBABILITY

The fault probability for an idealized periodic array of multiple patterns has been presented elsewhere [5, 8], and Figure 2.8 is a representation of an arbitrary defect size-probability density function $S(x)$, fault probability kernel $K(x - w)$, and their product $\phi(x)$. The area under $\phi(x)$ is the average fault probability Φ.

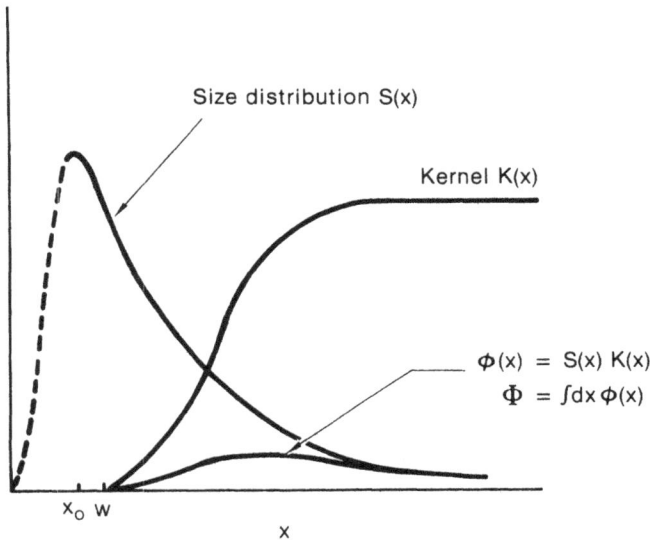

Figure 2.8 Size distribution, fault probability kernel, and fault probability. $S(x)$ is the defect size probability density function, $K(x - w)$ is the fault probability kernel, $\phi(x)$ is the product of $S(x)$ and $K(x - w)$, and Φ, the area under $\phi(x)$, is the average fault probability.

For realistic pattern geometries, the analytical treatment gets rapidly more complicated because actual product patterns are not all the same shape and size and do not have a simple periodicity. Furthermore, actual defects are not circular, tending more often to be like filaments of irregular shape. An analytical treatment such as that presented in this chapter can help the reader to understand the concepts more readily, but simulation methods [9–11] are needed to be able to determine the defect-sensitive area and the fault probability for actual products. Although a proper discussion and treatment of such methods is beyond the scope of this introductory text, the essential concepts are straightforward.

To represent a defect, a spot of arbitrary dimensions generated by a random number generator is placed at an arbitrary location on a computer-screen layout of a repetitive unit of the chip. If this spot breaks a pattern or connects two patterns, the appropriate fault count is recorded. This is done a large number of times for each defect of different size, and the ratio of the number of faults to the number of tries is taken to be the fault probability. The choice of how many defects of a given size to select depends on the assumptions made about the size distribution of the actual defects, assumptions that also affect the value of the fault probability and the defect-sensitive area, as will be shown in the next chapter.

2.7 REFERENCES

[1]   A.V. Ferris-Prabhu, "Critical Area for Rectangular Patterns," *IBM Technical Report* TR19.90199, 1981.

[2]   A.V. Ferris-Prabhu, "Fault Probability and Critical Area for VLSI Yield Projections," *IBM Technical Report* TR19.90200, 1982.

[3]   C.H. Stapper, "Modeling of Integrated Circuit Defect Sensitivities," *IBM Journal of Research and Development*, Vol. 27, 1983, pp. 549–557.

[4]   A.V. Ferris-Prabhu, "Computation of the Critical Area in Semiconductor Yield Theory," in *Proceedings of the European Conference on Electronic Design Automation (EDA84)*, University of Warwick, U.K., Publication 232, 1984, pp. 171–173.

[5]   A.V. Ferris-Prabhu, "Modeling the Critical Area in Yield Forecasts," *IEEE Journal of Solid State Circuits*, Vol. SC-20, No. 4, August 1985, pp. 874–878.

[6]   A.V. Ferris-Prabhu, "Role of Defect Size Distribution in Yield Modeling," *IEEE Transactions on Electron Devices*, Vol. ED-32, No. 9, September 1985, pp. 1727–1736.

[7]   A.V. Ferris-Prabhu, "Defects, Faults and Semiconductor Device Yield," *Defects and Fault Tolerance in VLSI Systems*, Vol. I, I. Koren, ed., Plenum Press, New York, 1989, pp. 129–137.

[8]   A.V. Ferris-Prabhu, "Yield Implications and Scaling Laws for Submicrometer Devices," *IEEE Transactions on Semiconductor Manufacturing*, Vol. 1, No. 2, May 1988, pp. 49–61.

[9]   J.M. Hammersley and D.C. Handscomb, *Monte Carlo Methods*, Methuen & Co., Ltd, 1964. Distribution in the U.S. by Barnes & Noble, Inc.

[10]   D.M.H. Walker, "Critical Area Analysis," *SRC Publication C91611*, Carnegie-Mellon University, August 1991.

[11]   D.M.H. Walker, *Yield Simulation for Integrated Circuits*, Kluwer Academic Publishers, 1987. This text contains many additional references.

Chapter 3
Effect of Defect Sizes on Fault Probability

3.1 INTRODUCTION

The previous chapter has shown that both the defect-sensitive area and the probability that a defect will be fatal depend on the pattern geometry, which determines the fault probability kernel, as well as the defect size-probability density function, here referred to for the sake of brevity as the size distribution. This chapter will assume a power-law dependence for the distribution of defect sizes and will examine the effect of changes in the assumptions on the magnitude of the defect-sensitive area and the fault probability.

3.2 DEFECT SIZE DISTRIBUTIONS

Major difficulties in obtaining the defect size distribution are that of determining a priori when a defect will be potentially fatal, counting all such defects, and then finding a way of characterizing their size. Any deviation from design can be considered a defect, but not all need be potentially fatal. Most fabrication facilities have visual inspection techniques after many, if not all, operations during which defects like pits, scratches, unlifted metal, unremoved photoresist, and other foreign materials that may result in a missing or an extra pattern are detected on a sample of wafers. However, it has been observed that there is little if any correlation between the yield predicted on the basis of the defects found during such observations and the yield determined electrically at wafer final test. In practice, test structures are designed to be sensitive to one or more different types of defects that, experience suggests, will occur. These structures usually consist of labyrinthine patterns arranged so that the "capture" of a defect is manifest by the pattern failing an electrical conductivity test. Many structures and methods [1–6] have been proposed, but for all of them it is, in essence, the yield that is determined, after which a physical failure analysis helps determine the type of defect causing the fail. Defects escaping capture do not result in failure and so tend not to be detected. Also, defects occur in a variety of shapes, and often it is the orientation of a defect as much as its size that determines whether it will be fatal. In consequence, neither the spatial nor the size-probability density function for defects occurring during semiconductor device fabrication is known with certainty. However, all existing data show that the number of defects larger than the threshold of resolution tends to decrease as some inverse power of their size, with no critical size reported at which the number peaks. However, it does not appear meaningful to consider that there is an infinite number of defects of vanishing size, so some workers [7, 8] have assumed that there is some critical size at which the probability density function peaks, and then decreases for smaller sizes. Such an assumption is not unreasonable, particularly if the critical size is chosen to be smaller than any present or reasonably foreseeable minimum device dimension, because the presence of defects smaller than the minimum device dimension cannot break a pattern or connect two adjacent patterns.

There are many functions that satisfy the requirement of peaking and going to zero on either side of the peak. They include the normal, lognormal, and Rayleigh density functions. Many of them are difficult to treat analytically, and given the uncertainty of the exact form, simulation methods to handle them may not be worthwhile. In the discussion that follows, it will be assumed that the size density function can be described by a power law for defects smaller than the critical size and by an inverse power law for defects larger than the critical size, and the effect of the assumed value of the power on the fault probability will be examined.

3.3 POWER-LAW ASSUMPTION

Consider the size density function, which is described in more detail elsewhere [9–10], defined by

$$S(x) = c_q x^q, \quad 0 \le x \le x_0 \tag{3.1a}$$

$$S(x) = c_p / x^p, \quad 0 \le x_0 \le x_M \tag{3.1b}$$

The function $S(x)$ is continuous at x_0, where $x_0 \ll w$ is the defect size at which the density function peaks, w is the minimum-dimension feature size, and x_M is the maximum size defect expected to occur in the area within which the pattern lies.

The requirement that 100% of all the defects lie within the interval $[0, x_M]$ is contained in the expression

$$1 = \int_0^{x_M} dx S(x) = \int_0^{x_0} dx c_q x^q + \int_{x_0}^{x_M} dx c_p / x^p \tag{3.2}$$

from which the constants are found to be

$$c_q = (q + 1)/x_0^{q+1}[1 - \ln(x_0/x_M)] \tag{3.3a}$$

$$c_p = (q + 1)/[1 - \ln(x_0/x_M)], \quad p = 1 \tag{3.3b}$$

and

$$c_q = (p - 1)(q + 1)/x_0^{q+1}$$
$$\{(p - 1) + (q + 1)[1 - (x_0/x_M)^{p-1}]\} \tag{3.4a}$$

$$c_p = (p - 1)(q + 1)x_0^{p-1}/$$
$$\{(p - 1) + (q + 1)[1 - (x_0/x_M)^{p-1}]\}, \quad p \neq 1 \tag{3.4b}$$

For these choices of $S(x)$, the single-pattern fault probability is

$$\Phi = (c_p/W) \times [(x_M - w) + \ln(w/x_M)], \quad p = 1 \tag{3.5a}$$

$$\Phi = (c_p/W) \times [(w/x_M - 1) - \ln(w/x_M)], \quad p = 2 \tag{3.5b}$$

$$\Phi = (c_p/W) \times f(w, x_M) \tag{3.5c}$$

where

$$f(w, x_M) = \frac{[1 - (w/x_M)^{p-2}] + (p - 2)/(p - 1)[1 - (w/x_M)^{p-1}]}{(p - 1)w^{p-2}}, \tag{3.5d}$$
$$p \geq 3$$

When the largest size defect x_M is much larger than the line width w, the fault probability is given by

$$\Phi \simeq \frac{1}{w^{p-1}}, \quad w \ll x_M \tag{3.5e}$$

with the average number of faults per chip being

$$\lambda = N_D \Phi \tag{3.6}$$

Although the magnitude of the effect may not readily be evident, it is clear from Equation (3.5) that the magnitude of the fault probability is distinctly affected by the choices made for p and q, and it has been reported [9] that when evaluating the fault probability and critical area, it pays to mind one's p's and q's.

A suggested [7] choice of power-law defect size-probability density function with $p = 3$ and $q = 1$ is shown in Figure 3.1.

A density function such as this, which increases linearly till the defect size reaches a critical value x_0 and then decreases as the inverse cube of the defect size, has interesting features. There are as many defects smaller than the critical size as there are defects that are larger. Furthermore, it has been shown [7] that, with such a choice, the number of faults per chip stays the same when the minimum feature dimension is reduced and the chip area reduced proportionately. Consequently such a size distribution function is yield neutral.

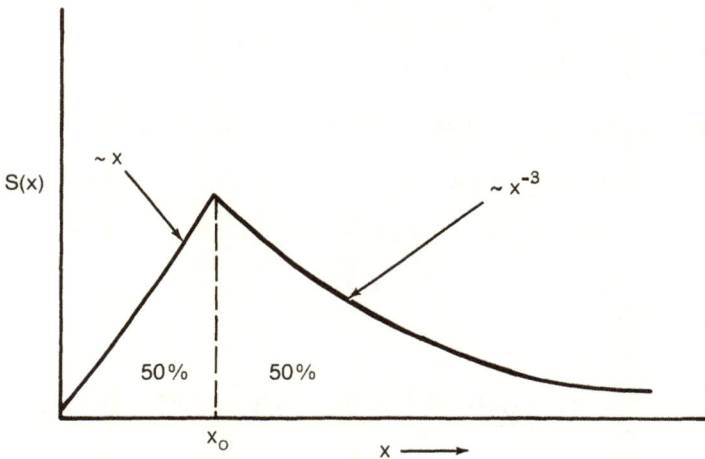

Figure 3.1 Defect size-probability density function with $q = 1$ and $p = 3$.

3.4 EFFECT OF DIFFERENT CHOICES OF POWER LAW

The effect of different choices of power law on the fault probability, defect-sensitive area, and number of faults [8–10] can be conveniently examined by comparing their values for different values of q and p to their value when $q = 1$ and $p = 3$; that is, by evaluating the ratio r, defined below, which has the same value for the relative fault probability, relative critical area (or relative defect-sensitive area), and the relative number of faults.

$$r = \Phi_{pq}/\Phi_{31} = A_{pq}/A_{31} = \lambda_{pq}/\lambda_{31} \tag{3.7}$$

where the fault probability is given by Equations (3.4) and (3.5), the defect-sensitive area is the product of the chip area and the fault probability, and the number of faults per chip is given by Equation (3.6).

Assuming a value of $x_0 = 0.1$ μm for the defect size at which the density function peaks, $w = 1$ μm for the minimum design dimension, $x_M = 10$ μm for the largest size defect of this type that, experience suggests, is likely to occur, and a chip of dimensions $X = Y = 1$ mm, using Equations (3.3) to (3.6) leads to the results shown in Table 3.1.

The choice of size distribution clearly affects the magnitude of the critical area, the fault probability, and the number of faults per chip, and this in turn affects the value of the predicted yield. With $q = 1$, the value of these quantities at $p = 2$ is about twelve times the value at $p = 3$, whereas at $p = 4$ the value is about one twelfth the value at $p = 3$. As stated earlier, with $q = 1$ and $p = 3$, the number of defects larger than the critical size is half the total number. This makes for very convenient scaling whenever it becomes necessary, as the resolution increases, to reset the value of the critical defect size. But more useful is the fact that a value of $p = 3$ is effectively yield neutral, in that if the chip size is reduced in the same proportion as the minimum design feature, there is no effect on the predicted yield. This feature makes it easy to evaluate and compare the yield of products with different size chips and ground rules without having to compute the defect-sensitive area and the fault probability in detail each time, using simple scaling laws to address the change in their magnitudes. However, although it has not been proven to be wrong, there is no physical law that requires the defect size distribution to vary as the inverse cube of defect size, so determination of the exact behavior is a fruitful area for further research.

Table 3.1
Relative Fault Probability, Critical Area, and Faults per Chip

p $\downarrow$	$q \rightarrow$ $r \searrow$	1	2	3	4
1		321.75	482.56	643.5	804.58
2		12.66	14.14	15.22	15.89
3		1.00	1.2	1.3	1.4
4		0.089	0.112	0.126	0.133

3.5 REFERENCES

[1]   G. Carver, L. Linholm, and T. Russell, "Use of Microelectronic Test Structures to Characterise IC Materials, Processes, and Processing Equipment," *Solid State Technology,* Vol. 23, No. 9, September 1980, pp. 85–92.

[2]   M. Buehler and L. Linholm, "Role of Test Chips in Coordinating Logic and Circuit Design and Layout Aids for VLSI," *Solid State Technology,* Vol. 24, No. 9, September 1981, pp. 68–73.

[3]   W. Maly *et al.,* "Systematic Characterization of Physical Defects for Fault Analysis of MOS IC Cells," *IEEE International Test Conference Proceedings,* 1984, pp. 390–399.

[4]   M.A. Mitchell, "Defect Test Structures for Characterization of VLSI Technology," *Solid State Technology,* Vol. 28, No. 5, May 1985, pp. 207–213.

[5]   W. Maly *et al.,* "Double-Bridge Test Structure for the Evaluation of Type, Size and Density of Spot Defects," *Research Report CMUCAD-87-2,* SRC-CMU Center for Computer-Aided Design, Carnegie-Mellon University, 1987.

[6]   W. Maly *et al.,* "Characterization of Type, Size and Density of Spot Defects in the Metallization Layer," in *Yield Modeling and Fault Tolerance in VLSI,* W. Moore, W. Maly, and A. J. Strojwas, eds., Adam Hilger, 1988.

[7]   C.H. Stapper, "Modeling of Integrated Circuit Defect Sensitivities," *IBM Journal of Research and Development,* Vol. 27, 1983, pp. 549–557.

[8]   A.V. Ferris-Prabhu, "Defect Size Variations and Their Effect on the Critical Area of VLSI Devices," *IEEE Journal of Solid State Circuits,* Vol. SC-20, No. 4, August 1985, pp. 878–880.

[9]   A.V. Ferris-Prabhu, "Role of Defect Size Distribution in Yield Modeling," *IEEE Transactions on Electron Devices,* Vol. ED-32, No. 9, September 1985, pp. 1727–1736.

[10]   A.V. Ferris-Prabhu, "Yield Implications and Scaling Laws for Submicrometer Devices," *IEEE Transactions on Semiconductor Manufacturing,* Vol. 1, No. 2, May 1988, pp. 49–61.

Chapter 4
Counting Techniques

4.1 INTRODUCTION

Device yield has been defined as the fraction of chips on a wafer that is functional. This fraction is usually averaged over some, many, or all wafers manufactured in a particular time period. Device yield can also be defined as the probability of obtaining a chip that has no fatal defects. Semiconductor yield prediction therefore requires calculating the probability of finding a particular state (a chip with no fault), out of all possible states (chips with zero, one, two, or more faults) when the events (faults) are distributed over the states according to some distribution law. In effect, then, calculating the probability involves the use of a counting technique appropriate to the laws governing the ways in which the events are distributed.

In this chapter we will show how to calculate the probability of finding a state with a particular event when the events are distributed according to the Binomial, Poisson, Maxwell-Boltzmann, Bose-Einstein, and Fermi-Dirac distributions. It is emphasized that expressions for the probability of an event occurring are mathematical representations of what we think are the physical laws underlying their occurrence. If prediction does not agree with reality, it may be because the mathematical representation is not valid in certain cases, or because all the underlying physical factors are not known, or their relationships to each other are not fully understood. In such cases there is no alternative to gathering more data, since the predictive capability of a theory depends on how well it is based on actual data.

4.2 BERNOULLI TRIALS

Bernoulli trials are a set of trials for each of which there are only two possible outcomes, whose probability of occurrence remains the same throughout the trials. It should be noted that the model of Bernoulli trials is a theoretical one and may not be in agreement with a particular physical situation. When that occurs, the model will need to be modified to reflect the physical reality more correctly.

An example of a Bernoulli trial is that of tossing a coin. When a coin is tossed, if a person waits long enough for it to fall on one side or the other, there can be only one of two outcomes: heads or tails. And if the coin is tossed repeatedly in the same manner, the likelihood of getting a head or a tail remains the same each time. A semiconductor chip, however, can have any number of faults on it, from zero upward, and there is no a priori reason for the occurrence of a fatal defect to be as likely on one chip as on another, or for the probability of occurrence to be unaffected by the prior occurrence of another and so to stay the same throughout. Nevertheless, for our purposes, there are only two outcomes of interest: the states with one or more faults and the state with no fault, and we are particularly interested in obtaining an expression for the probability of occurrence of the latter.

So it is instructional to use the example of coin tossing as a starting point to obtain expressions for the probability of occurrence of an event of interest, when the events are distributed with equal probability according to different distribution laws or statistics. In the end, only experience can tell whether the distribution law being used to describe a particular physical situation does so correctly.

The next two sections will develop the binomial distribution and the Poisson approximation to it following the well-known treatment attributable to Feller [1].

4.3 THE BINOMIAL DISTRIBUTION

In the case of coin tossing, if the coin is true, each outcome is equally likely. But in a Bernoulli trial, it is not necessary for the probability of each of the two outcomes to be the same. If the probability of one outcome (getting a head) is written as p, and the probability of the other (getting a tail) is written as q, then, since after a toss one *must* get *either* a head *or* a tail, it is evident that

$$p + q = 1 \tag{4.1}$$

The same holds true even after N tosses: the probability of getting *either* at least one head, *or* at least one tail, is unity. Thus,

$$(p + q)^N = 1 \tag{4.2}$$

Noting that q can also be written as $1 - p$, the left-hand side of Equation (4.2) can be expanded to give

$$(p + q)^N = \sum_{k=0}^{N} \frac{N!}{k!(N - k)!} \times p^k (1 - p)^{N-k} \tag{4.3}$$

which is known as the binomial distribution.

In terms of our coin tossing example, where either a single coin is tossed N times or N true coins are each tossed once, each term in the right-hand side of this equation gives the probability that k heads will be obtained, where k can assume any integer value from 0 to N. In particular, the probability of obtaining a total of n heads, and $(N - n)$ tails is

$$P(n; N, p) = \frac{N!}{n!(N - n)!} \times p^n(1 - p)^{N-n} \tag{4.4}$$

4.4 THE POISSON DISTRIBUTION

In examples like that of coin tossing, where the number of trials N is large, and the probability p is small, the product of these terms

$$\lambda = Np \tag{4.5}$$

is often of moderate size. In such a case, the probability $P(0; N, p)$ of obtaining *no* heads can be written as $(1 - p)^N$, or, from Equation (4.5),

$$P(0; N, p) = \left(1 - \frac{\lambda}{N}\right)^N \tag{4.6}$$

By taking the logarithm of both sides and making use of Taylor's expansion, Equation (4.6) can be rewritten as

$$\ln P(0; N, p) = N\ln\left(1 - \frac{\lambda}{N}\right) = -\lambda - \frac{\lambda^2}{2N} - \ldots \tag{4.7}$$

When N is very large, the second and successive terms become negligibly small, enabling Equation (4.7) to be rewritten as

$$P(0; N, p) \simeq e^{-\lambda} \tag{4.8}$$

which is the probability of not getting a head.

Dividing the probability $P(n; N, p)$ by the probability $P(n - 1; N, p)$ and using Equations (4.4) and (4.6), it is seen that

$$\frac{P(n; N, p)}{P(n - 1; N, p)} = \frac{Np - (n - 1)p}{n(1 - p)} \simeq \frac{\lambda}{n} \tag{4.9}$$

when N is much larger than n, and p is small enough to be neglected.

From this we find that the probability of getting at least one head is

$$P(1; N, p) \simeq \frac{\lambda}{1!} P(0; N, p) = \lambda e^{-\lambda} \qquad (4.10)$$

the probability of getting at least two heads is

$$P(2; N, p) \simeq \frac{\lambda^2}{2!} e^{-\lambda} \qquad (4.11)$$

and, in general,

$$P(n; N, p) \simeq \frac{\lambda^n}{n!} e^{-\lambda} \qquad (4.12)$$

This is the famous Poisson approximation to the binomial distribution.

If a number of events occur randomly in adjacent, nonoverlapping regions of area A, and if the events that occur in one region in no way affect or are affected by the events in any other of the regions, then the probability of obtaining exactly n events in one of the given regions is

$$P(n; \lambda) = \frac{\lambda^n}{n!} e^{-\lambda} \qquad (4.13)$$

If the region is taken to be a chip, and λ is interpreted as the average number of events (i.e., fatal defects per chip), then the probability of obtaining a chip with no fatal defects, the yield, can be written as

$$P(0; \lambda) = e^{-\lambda} \qquad (4.14)$$

This equation is based upon the Poisson distribution and assumes that an event (i.e., the occurrence of a fatal defect) is equally likely to occur on one chip as on another, and is not affected by whether or not a fatal defect has already occurred on a given chip. As we shall see later, this is not always the case.

4.5 THE MAXWELL-BOLTZMANN DISTRIBUTION

Another way in which events can be distributed is governed by the Maxwell-Boltzmann statistics. According to this law, the number of ways in which m independent and distinguishable events, each with the same probability of occurrence, can be distributed over n states is

$$N(m, n) = n^m \qquad (4.15)$$

Suppose there are m distinguishable defects that can be distributed over a wafer of total surface area S, in any one of n different regions, each of area A, with m being much less than n. In terms of the defect density D and the area of a region A or of the total wafer surface area S, the total number of defects can be written as

$$m = nAD = SD = n\lambda \qquad (4.16)$$

where

$$AD = \lambda \qquad (4.17)$$

The number of ways in which the m defects can be distributed among all n regions is given by Equation (4.15), so if one of the n regions is kept free of defects, the number of ways the m defects can be distributed among the remaining $(n - 1)$ regions is

$$N(m, n - 1) = (n - 1)^m \qquad (4.18)$$

The probability of finding a region free of defects (i.e., the yield) is given by the ratio

$$\frac{N(m, n - 1)}{N(m, n)} = \left(1 - \frac{1}{n}\right)^m = \left(1 - \frac{A}{S}\right)^{SD} \qquad (4.19)$$

Equation (4.19) in a slightly different form was first suggested by J. Wallmark [2].

When Equation (4.19) is expanded and written in terms of λ, it reduces to the Poisson equation

$$1 - \lambda + \frac{\lambda^2}{2!} - \ldots = e^{-\lambda} \qquad (4.20)$$

as n increases without bounds. This derivation was suggested by Price [3].

4.6 THE BOSE-EINSTEIN DISTRIBUTION

Although defects are in principle distinguishable, there are some defects, such as pinholes in the dielectric, that are virtually indistinguishable from each other. Being practically points, there is effectively no limit on the number of them that can occur in a given region. Such a situation is described by the Bose-Einstein statistics, according to which the number of ways in which m indistinguishable particles (defects) can be distributed with equal probability among n states (chips), with no restriction on how many particles can occupy a state, is given by the expression

$$N(m, n) = \frac{(m + n - 1)!}{m!(n - 1)!} \tag{4.21}$$

The number of ways in which the same m defects can occur in $(n - 1)$ regions is

$$N(m, n - 1) = \frac{(m + n - 2)!}{(m - 1)!(n - 1)!} \tag{4.22}$$

from which the probability of finding a single region free of defects (i.e., the yield) is given [3] by the ratio

$$\frac{N(m, n - 1)}{N(m, n)} = \frac{n - 1}{n - 1 + m} \approx \frac{1}{1 + \lambda} \tag{4.23}$$

where, as before, $m = nAD = n\lambda$, and n increases without bounds.

4.7 THE FERMI-DIRAC DISTRIBUTION

The Fermi-Dirac statistics give the number of ways in which m indistinguishable particles (defects) can be distributed with equal probability among n states, with the constraint that a state can be occupied by at most one particle. If there are m particles that are governed by this law, the number of ways in which they can be distributed among n states is

$$N(m, n) = \frac{n!}{m!(n - m)!} \tag{4.24}$$

The number of ways in which the same m particles can be distributed over $(n - 1)$ states is

$$N(m, n - 1) = \frac{(n - 1)!}{m!(n - m - 1)!}$$ (4.25)

As before, the probability of finding a state with no particle in it is given by the ratio

$$\frac{N(m, n - 1)}{N(m, n)} = 1 - \frac{m}{n} \simeq 1 - \lambda$$ (4.26)

There is no physical reason why not more than one defect of the same kind can occur on a chip, so this expression is not suitable for use as a yield equation. However, in a mature product with a high yield (i.e., a very low number of faults per chip), the Poisson equation $e^{-\lambda}$ can be approximated by the terms $(1 - \lambda)$. This is the mathematical expression of the physical fact that when the number of faults per chip is very small, the probability of two of them occurring on the same chip is virtually zero. Nevertheless, the Fermi-Dirac statistics do not apply to semiconductor defects, and this distribution law should not be used to try to derive a yield equation.

4.8 REFERENCES

[1]   W. Feller, *An Introduction to Probability Theory and its Applications*, Vol. I, 2nd edition, John Wiley, 1957. A more recent text on the subject is by A. Papoulis, *Probability, Random Variables, and Stochastic Processes*, 2nd edition, McGraw-Hill, 1984. Another way of obtaining the Poisson Distribution from the binomial distribution is given in M.R. Spiegel, *Schaum's Outline Series, Theory and Problems of Probability and Statistics*, p. 129.

[2]   J. Wallmark, "Design Considerations for Integrated Electronic Devices," *Proceedings of the IRE*, Vol. 48, No. 3, March 1960, pp. 293–300.

[3]   J.E. Price, "A New Look at Yield of Integrated Circuits," *Proceedings of the IEEE*, Vol. 58, No. 8, August 1970, pp. 1290–1291.

Chapter 5
Yield Equations

5.1 INTRODUCTION

The previous chapter discussed the number of ways in which m events could be distributed over n states according to different distribution laws. The probability of finding a state (chip) with no event (defect) was equated to the yield. This chapter starts by assuming that defects are distributed with equal probability over the wafer, leading to the Poisson yield equation. But, in fact, the defect density is not constant over the wafer, so a weighting function that describes the distribution of defect densities will be introduced into the Poisson equation, which is then integrated over all defect densities. This chapter also considers a few different defect density weight functions and shows their resulting yield equations.

5.2 SPATIAL DISTRIBUTION OF DEFECTS

We have seen that, during the fabrication of a device, defects of different types can occur: gross defects, which almost always affect device yield; parametric defects, which infrequently affect yield, although they can affect reliability; and random defects, which have a probability of becoming fatal and thus affecting the yield. Chapter 2 showed how to calculate the probability that a random point defect or one with spatial extent would become fatal, and Chapter 3 showed that the defect size-probability density function affected this probability. The fault probability is the probability that a random defect will become a fault, given the probability that a random defect occurs in a given region. The probability that a random defect occurs in a given region is determined by the defect spatial probability density function, or, equivalently, the distribution of defect densities over the wafer. Thus, the yield depends on both the size distribution and on the spatial distribution of defects [1].

   The simplest assumption to make is that a defect has an equal chance of occurring in one location as at another; that is, the defect density is constant across the wafer. Some other assumptions about the defect density distribution are that it is Gaussian, uniform, or exponential.

5.3 CONSTANT DEFECT DENSITY YIELD EQUATION

If the probability of obtaining a fatal defect is the same on one chip as on another, then, as shown in the previous chapter, the probability of finding a chip with n fatal defects is given by

$$p(n, \lambda) = \frac{\lambda^n e^{-\lambda}}{n!} \tag{5.1}$$

where $\lambda = AD$ is the average number of fatal defects per chip, D is the density of potentially fatal defects, and A is the defect-sensitive area.

The probability of finding a chip with no defects on it (i.e., the yield) is

$$Y = p(0, \lambda) = e^{-\lambda} \tag{5.2}$$

which is, of course, the well-known Poisson yield equation.

In general, if the defect density is not constant, but varies over the wafer, governed by its own probability density function $f(D)$, the probability of finding n defects on a chip is given by

$$p(n, AD) = \int dD\, f(D)\, \frac{(AD)^n e^{-AD}}{n!} \tag{5.3}$$

which leads to the yield expression [2]

$$Y = p(0, AD) = \int dD\, f(D) e^{-AD} \tag{5.4}$$

where the average number of faults λ is replaced by the product of the defect-sensitive area A and the (potentially fatal) defect density D.

The assumption that the defect density is constant can be written as

$$f(D) = \delta(D - D_0) \tag{5.5}$$

where the Dirac delta function $\delta(D - D_0)$ requires the function $f(D)$ to vanish everywhere except at $D = D_0$. Inserting Equation (5.5) into Equation (5.3) gives

$$Y = \int dD\, \delta(D - D_0) e^{-AD} = e^{-AD_0} \tag{5.6}$$

which again gives the Poisson yield equation, as it should. The Poisson yield equation has been widely reported to predict lower yields than that subsequently observed.

5.4 SIMPSON DISTRIBUTION YIELD EQUATION

The form of $f(D)$ is not known, and different forms have been proposed based on data from test structures and interpretation of actual yield data. A not unreasonable assumption is that $f(D)$ is Gaussian, which, if approximated for mathematical simplicity by a Simpson (i.e., triangular) distribution of the form

$$f(D) = \frac{D}{D_0^2}, \quad 0 \le D \le D_0 \tag{5.7a}$$

$$f(D) = \frac{1}{D_0}\left(2 - \frac{D}{D_0}\right), \quad D_0 \le D \le 2D_0 \tag{5.7b}$$

gives the yield equation

$$Y = \int_0^{D_0} dD\, \frac{D}{D_0^2} e^{-AD} + \int_{D_0}^{2D_0} dD\, \frac{1}{D_0}\left(2 - \frac{D}{D_0}\right)e^{-AD} \tag{5.8a}$$

or

$$Y = \left(\frac{1 - e^{-AD_0}}{AD_0}\right)^2 \tag{5.8b}$$

which is known as Murphy's yield equation. The predictions of this equation have been reported [2] to agree tolerably well with actual yields.

5.5 UNIFORM DISTRIBUTION YIELD EQUATION

If the density function $f(D)$ is assumed to be uniform over the interval $[0, 2D_0]$ such that

$$f(D) = \frac{1}{2D_0}, \quad 0 \le D \le 2D_0 \tag{5.9}$$

the yield is given by

$$Y = \int_0^{2D_0} dD\, \frac{1}{2D_0} e^{-AD} = \left(\frac{1 - e^{-2AD_0}}{2AD_0}\right) \tag{5.10}$$

This yield equation has been reported [2] to predict yields higher than are actually observed.

5.6 HALF-GAUSSIAN DISTRIBUTION YIELD EQUATION

If the function $f(D)$ can be approximated by a half Gaussian of the form

$$f(D) = \frac{2}{\pi D_0} e^{-(D/D_0 \sqrt{\pi})^2}, \quad D \geq 0 \tag{5.11}$$

the yield equation

$$Y = \int_0^\infty dD \, \frac{2}{\pi D_0} e^{-(D/D_0 \sqrt{\pi})^2} e^{-AD} \tag{5.12}$$

can be evaluated to give

$$Y = e^{\pi(AD_0/2)^2} \text{erf}\left(\sqrt{\pi \frac{AD_0}{2}}\right) \tag{5.13}$$

This equation was first published by Stapper who has reported [3] that it has since been superseded by other yield equations that are easier to use and agree better with actual data.

5.7 EXPONENTIAL DISTRIBUTION YIELD EQUATION

If it is assumed [4, 5] that the defect density is exponentially distributed, that is,

$$f(D) = \frac{e^{-D/D_0}}{D_0} \tag{5.14}$$

the yield equation

$$Y = \int_0^\infty dD \, \frac{e^{-D/D_0}}{D_0} e^{-AD} \tag{5.15}$$

is obtained, which integrates to

$$Y = \frac{1}{(1 + AD_0)} \tag{5.16}$$

which is identical to the yield expression obtained in the previous chapter, assuming that the occurrence of defects is governed by the Bose-Einstein statistics.

This equation is sometimes referred to as Seeds' equation, after Seeds, who first introduced and later modified it to agree with his data [6], and sometimes referred to as Price's equation after Price [7], who also derived it, but from different considerations. Its predictions have been found to be higher than the actual yield.

If there are s independent mechanisms that generate a total of m defects with densities D_1, D_2, ... D_s, it is possible to invoke the method shown in the previous chapter for Bose-Einstein distributed defects, and show that the probability of finding a chip with no defects is

$$\frac{(1 - 1/n)}{(1 + AD_1 - 1/n)(1 + AD_2 - 1/n) \ldots (1 + AD_s - 1/n)}$$

which, as the number of states n increases without bounds, gives the yield equation

$$Y = \prod_{k=1}^{s} \frac{1}{1 + AD_k} \tag{5.17}$$

This result, due to Price, is essentially the same as assuming, as Seeds did, that during fabrication, the defects generated at each mask level can be described by the exponential distribution given in Equation (5.14). Many yield practitioners use Equation (5.17) because each term in the product can be interpreted as the yield of a particular set of fabrication steps associated with a given mask level.

5.8 ERLANG DISTRIBUTION YIELD EQUATION

It was assumed by Okabe [8] that each of the s mask levels has the same defect density D_0 and probability density function $e^{(-D/D_0)}/D_0$. Then the overall density function, known as Erlang's probability density function [9], is

$$f(D) = \frac{(s/D_0)^s}{(s - 1)!} e^{-sD/D_0} D^{s-1} \tag{5.18}$$

giving the yield equation

$$Y = \int_0^\infty dD \frac{(s/D_0)^s}{(s - 1)!} e^{-sD/D_0} D^{s-1} e^{-AD} = \frac{1}{(1 + AD_0/s)^s} \tag{5.19}$$

This equation may be called Okabe's yield equation. It has been reported [6] that its predictions do not agree well with data.

5.9 GAMMA DISTRIBUTION YIELD EQUATION

Another distribution proposed [10, 11] for $f(D)$, and based on the gamma function, can be written as

$$f(D) = \frac{1}{\Gamma(\alpha)B^\alpha} D^{\alpha-1} e^{-D/B} \tag{5.20}$$

where

$$\alpha = \frac{D_0^2}{\text{var}(D)}, \quad B = \frac{\text{var}(D)}{D_0}, \quad D_0 = \alpha B$$

and D_0 is the average defect density.

Equation (5.20) inserted into Equation (5.3) shows that the probability of obtaining a chip with n defects on it is

$$p(n, AD) = \frac{\Gamma(\alpha + n)}{n!\Gamma(\alpha)} \frac{(AD_0/\alpha)^n}{(1 + AD_0/\alpha)^{n+\alpha}} \tag{5.21}$$

from which the yield equation is given by

$$Y = \frac{1}{(1 + AD_0/\alpha)^\alpha}, \quad 1 \le \alpha \le \infty \tag{5.22}$$

This equation, which is structurally similar to Okabe's equation, though derived from different considerations, is referred to as the negative binomial yield equation. Its predictions have been reported [6, 10] to agree well with data.

The parameter α is related to the variance of the defect density D and is a measure of the extent to which the defects are clustered. When $\alpha = 1$, it reduces to the yield equation based on the Bose-Einstein distribution or, equally, the exponential distribution, implying a high degree of clustering. When $\alpha = \infty$, it passes over into the Poisson equation, implying no clustering. More physical insight is obtained if α is replaced by its reciprocal, β. The term β has been interpreted [12] as a coefficient coupling the occurrence of defects, or as the extent of deviation from Poisson statistics. When the coupling coefficient $\beta = 0$, there is no coupling of defects (i.e., no clustering), and a value of $\beta = 1$ implies no constraint on the number of defects clustered in the same location. This replacement allows Equation (5.22) to be rewritten as

$$Y = \frac{1}{(1 + \beta AD_0)^{1/\beta}}, \quad 0 \le \beta \le 1 \tag{5.23}$$

The negative binomial equation is a two-parameter equation; the second parameter β (also written as $1/\alpha$) gives it considerable flexibility. This distribution has been known for a long time [13, 14] and has been applied to problems in eugenics [15], psychology [16], ecology [17], and astrophysics [18], the common feature in these applications being dispersion of human characteristics, of plant species, or of galaxies. Therefore, it is well suited to studying the effects of dispersion of semiconductor defects on the yield and for use in semiconductor yield prediction.

5.10 DISCUSSION

There is a variety of other yield equations [6], but the ones described above are some of the better known expressions. They have been derived as modifications of the Poisson yield equation to accommodate the fact that defects do not occur with equal probability across the wafer, and require knowledge of the probability function $f(D)$ describing how the defect density varies across the wafer.

Unfortunately, there has not yet been demonstrated a method of determining $f(D)$ directly (i.e., without *inferring* it from the yield). In practice, a physically acceptable form for $f(D)$ is selected and a yield equation is derived and then applied to computing the yield of a product in production. If the equation gives yield values that match the actual yield data reasonably well, the choice of $f(D)$ that led to that yield equation cannot be said to be wrong. However, there is no à priori reason that the defect-generating mechanisms for one technology in a given period of its product cycle should be the same as that for the same product in a different period or for a different technology. Thus, a yield equation based upon a particular choice of $f(D)$, no matter how well it fits existing data, is subject to the possibility of not being applicable to another product. For these reasons it is often preferable to use a flexible model that can be easily modified to accommodate changes in the yield environment. The presence of the parameter β (or $1/\alpha$) both in the argument and as an exponent gives the negative binomial yield equation considerable flexibility. It can be used with data from previous products, and when yield data from early hardware is available, small adjustments can be made to the parameter β, leading to acceptable early predictions.

5.11 REFERENCES

[1]   A.V. Ferris-Prabhu, "Role of Defect Size Distribution in Yield Modeling," *IEEE Transactions on Electron Devices*, Vol. ED-32, No. 9, September 1985, pp. 1727–1736.

[2]   B.T. Murphy, "Cost-Size Optima of Monolithic Integrated Circuits," *Proceedings of the IEEE*, Vol. 52, No. 12, December 1964, pp. 1537–1545.

[3]   C.H. Stapper, "On Murphy's Yield Integral," *IEEE Transactions on Semiconductor Manufacturing*, Vol. 4, No. 4, November 1991, pp. 294–297.

[4]   R.B. Seeds, "Yield, Economic, and Logistic Models for Complex Digital Arrays," *IEEE International Convention Record, Part 6*, April 1967, pp. 61–66.

[5]   R.B. Seeds, "Yield and Cost Analysis of Bipolar LSI," *IEEE International Electron Devices Meeting*, October 1967, p. 12.

[6]   C.H. Stapper, "Fact and Fiction in Yield Modeling," *Microelectronics Journal*, Vol. 20, No. 1/2, Spring 1989, pp. 129–151.

[7]   J.E. Price, "A New Look at Yield of Integrated Circuits," *Proceedings of the IEEE*, Vol. 58, No. 8, August 1970, pp. 1290–1291.

[8]   T. Okabe, M. Nagata and S. Shimada, "Analysis on Yield of Integrated Circuits and a New Expression for the Yield," *Electrical Engineering in Japan*, Vol. 92, No. 6, December 1972, pp. 135–141.

[9]   P. Beckman, *Elements of Applied Probability Theory*, Harbrace Series in Electrical Engineering, Harcourt, Brace and World, 1968.

[10]   C.H. Stapper, "Defect Density Distribution for LSI Yield Calculations," *IEEE Transactions on Electron Devices (Correspondence)*, Vol. ED-20, July 1973, pp. 655–657.

[11]   T. Michalka, R. Varshney, and J. Meindl, "A Discussion of Yield Modeling with Defect Clustering, Circuit Repair and Circuit Redundancy," *IEEE Transations on Semiconductor Engineering*, Vol. 3, No. 3, August 1990, pp. 116–127.

[12]   A.V. Ferris-Prabhu, "Defects, Faults and Semiconductor Yield," *Defect and Fault Tolerance in VLSI Systems*, Vol. I, I. Koren, ed. Plenum Press, New York, 1989, pp. 129–137.

[13]   P.R. Montmort, "Essai d'analyse sur les jeus des hasards," 1714. Cited in J. Gurland, "Some Applications of the Negative Binomial and Other Contagious Distributions," *American Journal of Public Health*, Vol. 49, No. 10, 1959, pp. 1388–1399.

[14]   Student, "On the Error of Counting With a Haemocytometer," *Biometrika*, 1907, pp. 351–360.

[15]   R.A. Fisher, "The Negative Binomial Distribution," *Annals of Eugenics*, Vol. 11, 1941, pp. 182–187.

[16]   H.S. Sichel, "The Estimation of the Parameters of a Negative Binomial Distribution With Special Reference to Psychological Data," *Psychometrika*, Vol. 16, 1952, pp. 102–127.

[17]   G.E. Blackman, "Statistical and Ecological Studies in the Distribution of Species in Plant Communities. I. Dispersion as a Factor in the Study of Changes in Plant Populations," *Annals of Botany, London New Series*, Vol. 6, 1951, pp. 351–370.

[18]   J. Neyman and E.L. Scott, "A Theory of the Spatial Distribution of Galaxies," *Astrophysical Journal*, Vol. 116, 1952, pp. 144–163.

Chapter 6
Defect Density and Scaling Rules

6.1 INTRODUCTION

In this chapter we will examine two additional requirements of a yield equation, namely, that the net yield of a product equals the product of the yield of each of its layers, and that the average number of fatal defects per chip equals the sum over all layers of the average number of fatal defects per layer. We will also introduce various scaling rules for the defect density, and show how to satisfy these two requirements within the framework of the Poisson yield equation.

6.2 THE DEFECT DENSITY

The previous chapter showed that the spatial distribution of potentially fatal defects determines the choice of yield equation to use, which requires knowledge of λ, the average number of fatal defects per chip. As shown in an earlier chapter, the average number of fatal defects per chip depends on the size distribution of potentially fatal defects and their average density. Complete knowledge of the size and spatial distributions of all potentially fatal defects is therefore needed to be able to predict their effect on yield. It is clear that for a product yet to be built, there can naturally be no actual data available, so some sort of scaling to data from an existing relevant product is needed. However, even for an existing product it is of central importance to note that it is not practicable, even if it were possible, to determine directly the size and spatial distribution of all potentially fatal defects for at least two reasons. First, it is often not clear until a defect turns out to be fatal, that it was potentially fatal. Second, it is not possible to examine every chip for all possible departures from design and to determine their spatial and size distributions. In practice, a sample of chips that have failed test are physically failure analyzed, and through extensive delayering operations the physical cause of the fail that has been recorded electrically is determined, as well as the mask level at which that particular defect was introduced. This procedure on an ongoing basis is often supplemented by the use of test chips fabricated on the product chip wafers and designed to be susceptible to one or more particular defects. While this information is crucial to any analysis of semi-conductor yield, it is important to note that in almost all cases it is *yield* data that is recorded; that is, a particular electrical test is either passed or failed. Assumptions previously made about the size and spatial distributions of defects are used to select a yield equation, which is then inverted, making use of the experimentally deter-mined yield, to infer the average number of fatal defects. If a particular yield equa-tion $p(\lambda)$ is selected to describe the random defect-limited yield, and if Y_r is the random defect-limited yield component of the actually observed yield Y_{obs}, then the average number of fatal defects per chip is obtained from the relation

$$\lambda = p^{-1}(Y_r) \tag{6.1}$$

It is evident that the value of λ depends not only on the value of Y_r, but also on the yield equation chosen. The value of the average density of potentially fatal defects is then given by

$$D = \frac{\lambda}{A} = \frac{p^{-1}(Y_r)}{\Phi A_T} \tag{6.2}$$

where the defect-sensitive area A is the product of the total chip area and the fault probability, which itself depends on the assumptions made about the size distribution of potentially fatal defects.

The defect density, though an important quantity, is not itself measured di-rectly, and, as shown above, its value is inferred from yield measurements with assumptions about the spatial and size distributions of potentially fatal defects. Spe-cial test chips constructed to determine the size distribution of defects that can result in missing patterns or extra patterns have been reported, but, at best, they describe defect distributions on test chip wafers in an existing technology. So the expected defect density or faults per chip on products yet to be built in a future technology has to be scaled from the number of faults per chip of relevant product inferred from currently available yield data on these products. Before discussing scaling rules, however, a brief digression concerning yield models may be helpful.

6.3 YIELD MODELS

Yield models can be characterized in many ways. One way is to characterize them as either composite or layered. Composite yield models predict the yield based on the composite chip and the average number of faults of all types. Layered yield models predict the yield of each layer that was due to faults of each type that can affect that layer, and take their product to be the net yield. Consider the simplest type of yield equation, namely, the one that assumes that the probability of occurrence of a fatal defect is the same everywhere (i.e., the Poisson yield equation). Applying this equation to the composite yield model, if there is on average λ faults of all kinds per chip, the probability of finding a chip with zero defects (i.e., the yield) is

$$Y = p(0, \lambda) = e^{-\lambda} \tag{6.3}$$

Applying the Poisson equation to the layered model, if the average number of faults at the ith level is λ_i, the yield of the ith layer is

$$Y_i = p_i(0, \lambda_i) = e^{-\lambda_i} \tag{6.4}$$

and the net yield is

$$Y = \prod_{i=1}^{N} Y_i \tag{6.5}$$

where N is the number of layers.

The net yield should be the same no matter which model is used. This requirement leads to the result

$$\lambda = \sum_{i=1}^{N} \lambda_i \tag{6.6}$$

which says that the average of the sum of all faults is the sum over all layers of the average number of faults per layer. This leads to two conclusions. First, yield equations in which the product of the individual yields does not equal the net yield may not be the correct equation to use if a layered model to predict yield is used. Second, faults per chip, unlike defect densities, are dimensionless quantities, or pure numbers, and can be treated likewise.

Let us consider the implication of the first conclusion. If the defect density is constant across a wafer, the average number of defects per chip should increase linearly with chip area, and the Poisson yield equation should predict correct yields. In fact, neither happens. It has consistently been reported that the yield of chips larger than that of the product used as the reference is higher than that predicted by the Poisson equation, and examination of product chips as well as test chips has shown that the number of defects per unit area is not constant across the wafer. Defects have been found to occur in clusters with a noticeably higher incidence toward the periphery of the wafer [1, 2]. It is precisely to treat such situations that many different yield equations have been proposed, some of which were discussed in the preceding chapter. Several of them are quite complicated and none but the Poisson satisfies the requirement that the product of each of the individual mask level yields equals the net yield, or, equivalently, that the sum of the faults at each mask level add up to the total number of faults over all levels. This can be shown most readily for the negative binomial yield equation. Equating the net yield to the product of N individual mask level yields gives the relation

$$(1 + \beta\lambda)^{-1/\beta} = \prod_{i=1}^{N} [1 + (\beta_i\lambda_i)]^{-1/\beta_i} \tag{6.7}$$

The same value is often chosen for each of the cluster coefficients β_i and for β. Doing so and simplifying Equation (6.7) gives the result

$$\lambda = \sum_{i=1}^{N} \lambda_i + \beta \sum_{i=1}^{N} \sum_{j \neq i}^{N} \lambda_i\lambda_j + \cdots \tag{6.8}$$

which does not satisfy the requirement that the sum of the average number of faults at each level be equal to the average of the sum of all faults unless $\beta = 0$, in which case the negative binomial yield equation becomes identical to the Poisson yield equation. Indeed, if the relationship

$$f(\lambda) = \prod_{i=1}^{N} f(\lambda_i) \tag{6.9}$$

also requires that

$$\lambda = \sum_{i=1}^{N} \lambda_i \tag{6.10}$$

then the only functional relation that will satisfy this requirement is the exponential; that is, it is necessary to have

$$f(\lambda) \equiv e^{-\lambda} \tag{6.11}$$

This suggests that the reported discrepancies between the predictions of the Poisson model and actually observed yield may be due to an incorrect choice of the value to use for the average number of faults per chip (i.e., in the application of the equation rather than in its applicability).

In fact, if the number of faults per chip of a new product is scaled appropriately to the number of faults per chip inferred from the yield of a reference product, it is possible to satisfy the requirement that the average number of all faults be the same as the sum of the average number of faults at each mask level, and also possible to make reasonably correct yield predictions. The scaling rules needed to do so will be discussed next.

6.4 AREA SCALE FACTOR

The clearest indication that the defect density, or, equivalently, the average number of fatal defects per chip, is not constant across a wafer is the observation that the likelihood of finding a good chip decreases with the radial distance of the chip from the center of the wafer. From data that have been presented [1, 2] showing that the yield falls off steeply with the wafer radius, it has been found that the average number of faults per chip increases with the radius. This suggests that larger chips have a greater probability of having faults because they are located further from the center of the wafer than smaller chips. A good way to investigate the effect of the observed radial dependence of yield on the manner in which the average number of faults per chip scales with chip area is to examine the yield of memory products. Memory chips have a high degree of repetition, in that each chip consists of several identical array segments with common decode and peripheral circuitry. The yield of such chips can therefore be analyzed by treating them as "pseudochips" of one, two, three, or more array segments each. These pseudochips differ only in their area, so if the average number of faults per pseudochip is inferred from its yield, it should be possible to obtain the relationship between area and the average number of faults. The results of such a study [3] on a bipolar static random access memory (SRAM) chip with ten array segments has shown that the number of faults per pseudochip is given by

$$\lambda_N(\xi) = (e^{1/\xi} - 1) + cN \tag{6.12}$$

where c is an empirical constant related to the intrinsic defect density, N is the number of array segments in the pseudochip, and ξ is the radial distance from the periphery of the wafer to the outer vertex of the pseudochip. The ratio of faults on an N array segment pseudochip to those on a one array segment pseudochip is then given by

$$\lambda_N(\xi)/\lambda_1(\xi) = [N + \gamma(\xi)]/[1 + \gamma(\xi)] \tag{6.13}$$

with

$$\gamma(\xi) = (e^{1/\xi} - 1)/c \tag{6.14}$$

Noting that for pseudochips at the periphery, where $\xi = 0$,

$$\lambda_N(0)/\lambda_1(0) \to 1 \qquad (6.15a)$$

and for pseudochips at the center of the wafer where $\xi \to \infty$,

$$\lambda_N(\infty)/\lambda_1(\infty) \to N \qquad (6.15b)$$

suggests that Equation (6.13) can be written as

$$\lambda_N(\xi)/\lambda_1(\xi) = N^{1-b(\xi)}, \quad 0 \le b(\xi) \le 1 \qquad (6.16)$$

Because the area A_N of a pseudochip with N array segments is N times that of one with only one array segment, Equation (6.16) may be rewritten as

$$\lambda(A_N)/\lambda(A_1) = \alpha \qquad (6.17)$$

where the average area scaling factor α is given by

$$\alpha = \left(\frac{A_N}{A_1}\right)^{1-b}, \quad 0 \le b \le 1 \qquad (6.18)$$

and b is interpreted as a measure of the deviation of the defect density from constancy due to clustering. This measure is an empirical one and can be determined by comparing the average number of faults per test chip of different sizes as inferred from their yield. It has been reported [4] that for a wide range of products, a value of $b = 0.68$ appears to be valid, although other values have also been reported [5]. A value of $b = 0$ implies that defects are distributed with equal density over the wafer. Values of b greater than unity would imply that large chips have a lower average number of faults than smaller chips, but this has not yet been observed.

6.5 SENSITIVITY SCALE FACTOR

Another factor that affects the number of faults is the sensitivity of a product to defects. One such factor is clearly the minimum dimensions of the patterns. Even if the size distribution and average density of defects are the same, products with smaller design dimensions will be susceptible to smaller defects. Therefore, the defect-sensitive area, or, equivalently, the fault probability, will increase. To reflect this increase, the average number of faults obtained from the existing reference product will need to be scaled by a sensitivity factor ψ; that is,

$$\lambda_i = \psi \lambda_{e,i} \tag{6.19}$$

where λ_i is the number of faults on the ith layer of the new product, $\lambda_{e,i}$ is the number of faults on the corresponding layer of the existing product, and, in terms of the fault probabilities of existing and new products,

$$\psi = \frac{\Phi(w)}{\Phi(w_e)} \simeq \left(\frac{w_e}{w}\right)^{p-1} \tag{6.20}$$

where w and w_e are the minimum design dimensions on new and existing products, respectively.

If, as is often assumed, the defect size probability density function varies as the inverse cube of the size (i.e., $p = 3$), then

$$\psi \simeq \left(\frac{w_e}{w}\right)^{2} \tag{6.21}$$

A value of $p = 3$ is essentially yield neutral because it implies that the number of defects increases as the inverse square of their size. So if the chip area is reduced in the same proportion as the minimum design dimension, the increase in defect sensitivity is negated by the decrease in chip area. Thus, the yield, if there are no other changes, will be the same.

6.6 COMPLEXITY SCALE FACTOR

Because the total number of all faults is the sum of the number of faults at each mask level, the more mask levels there are for a given product, the greater the total number of faults will be. In a layered model this is accounted for by the use of as many mask level yields as there are mask levels. However, a particular mask level in one technology may be more (or less) complicated than the one to which it is being related in an existing technology. Consequently, more (or less) defects may be generated at that mask level. It is difficult to obtain a quantitative assessment for this additional complexity factor, and a subjective but not therefore unsubstantive assessment is often needed. Comparison of yield data with actual processes suggests that a factor of between 0.90 and 1.10, multiplying the number of faults obtained from the reference product mask level, provides reasonably accurate predictions. If $\lambda_{e,i}$ is the number of faults inferred from the yield of the ith level of an existing product, then the number of faults for that level of the new product can be expressed as

$$\lambda_i = \xi_i \lambda_{e,i}, \quad 0.9 \leq \xi_i \leq 1.10 \tag{6.22a}$$

with corresponding yield

$$Y_i = e^{-\lambda_i} \tag{6.22b}$$

6.7 POISSON YIELD EQUATION REVISITED

The area, sensitivity, and complexity scale factors may be combined to express the expected average number of faults λ_i on the ith layer of the new product, in terms of the number of faults $\lambda_{e,i}$ of the existing product to which it is being scaled, by the simple relationship

$$\lambda_i = \sigma_i \lambda_i \tag{6.23}$$

where

$$\sigma_i = \alpha_i \psi_i \xi_i \tag{6.24}$$

Since σ_i is merely a numerical multiplicative factor, it can be inserted into the Poisson yield equation to give for the yield of the ith mask level of the new product the expression

$$Y_i = e^{-\lambda_i} = e^{-\sigma_i \lambda_{e,i}} = Y_{e,i}^{\sigma_i} \tag{6.25}$$

and for the net yield

$$Y = e^{-\lambda} = e^{-\sigma \lambda_e} = Y_e^{\sigma} \tag{6.26}$$

where

$$\lambda = \sum_{i=1}^{N} \lambda_i \qquad (6.27)$$

and

$$\sigma = \sum_{i=1}^{N} \sigma_i \lambda_{e,i}/\lambda_e \qquad (6.28)$$

It can be readily verified that if

$$Y_e = \prod_{i=1}^{N} Y_{e,i} \qquad (6.29a)$$

and

$$\lambda_e = \sum_{i=1}^{N} \lambda_{e,i} \qquad (6.29b)$$

then

$$Y = \prod_{i=1}^{N} Y_i \qquad (6.30a)$$

and

$$\lambda = \sum_{i=1}^{N} \lambda_i \qquad (6.30b)$$

satisfying the requirement that the product of the individual yields equals the net yield, and that the sum of the average number of all faults equals the average number of faults per layer summed over all layers.

These useful results allow the Poisson equation to be used quite effectively to make yield predictions, as will be shown in the next chapter.

6.8 REFERENCES

[1]   T. Yanagawa, "Yield Degradation of Integrated Circuits Due to Spot Defects," *IEEE Transactions on Electron Devices*, Vol. ED-19, No. 2, 1972, pp. 190–197.

[2]   A.V. Ferris-Prabhu, L.D. Smith, H. Bonges, and J.K. Paulsen, "Radial Yield Variations in Semi-conductor Wafers," *IEEE Circuits and Devices Magazine*, Vol. 3, No. 2, March 1987, pp. 42–47.

[3]   A.V. Ferris-Prabhu and M. Retersdorf, "The Effect on Yield of Clustering and Radial Variations in Defect Density," in *Defect and Fault Tolerance in VLSI Systems*, Vol. 2, C. Stapper, ed., Plenum Press, 1990, pp. 63–73.

[4]   C.H. Stapper, "Fact and Fiction in Yield Modeling," *Microelectronics Journal*, Vol. 20, No. 1/2, Spring 1989, pp. 129–151.

[5]   A.V. Ferris-Prabhu, "A Cluster-Modified Poisson Model for Estimating Defect Density and Yield," *IEEE Transactions on Semiconductor Manufacturing*, Vol. 3, No. 2, May 1990, pp. 54–59.

Chapter 7
Yield Prediction

7.1 INTRODUCTION

This chapter shows how to predict yield for a new product by scaling yield data from an existing product to which it is being referenced. This chapter will also weave together several skeins of thought that were developed in earlier chapters to present a method of yield prediction in a manner that will enable it to be used for other cases as well.

7.2 COMPOSITE YIELD MODEL

Because the defect density or the average number of faults per chip is not known in advance, yield prediction for a new product must by necessity be scaled to information pertaining either to a relevant reference product currently being manufactured, or to special test structures fabricated in the technology of the proposed product, or both. Two sorts of information are needed: design information about the new and reference products such as the chip area, the number of mask levels, (and the minimum design dimension on each level) and yield information from the reference product so that the average number of faults due to each type of yield detractor can be evaluated for the reference product and then scaled to give their expected values for the new product. The way in which this may be done will be clearer if some of the salient features discussed in earlier chapters are repeated here in a more focused fashion.

If the yield prediction is to be made on the basis of a composite yield model, it suffices to know the chip area, number of mask levels and minimum dimension feature on both proposed and reference products, and the yield of the reference product. Yield data for the reference product is usually obtained by selecting, on an ongoing basis, a few wafers from each lot, testing them, and counting n_f, the number of chips that fail, out of n_T, the total number of chips tested. The net yield is then given by

$$Y_e = 1 - \frac{n_f}{n_T} \qquad (7.1)$$

where the subscript e is used to identify the existing reference product.

It does not matter which yield equation is used to infer the average number of faults, as long as it is used consistently. For simplicity, and also because it has been found to give good results if properly used, the Poisson yield equation will be used here, according to which the average number of faults of all types is

$$\lambda_e = -\ln\left(1 - \frac{n_f}{n_T}\right) \qquad (7.2)$$

If the areas of the existing and proposed products are A_e and A, the minimum feature dimensions w_e and w, and the number of mask levels N_e and N, respectively, the number of faults of all types expected on the new product is likely to be

$$\lambda = \sigma \times \lambda_e \qquad (7.3a)$$

where the overall scale factor is

$$\sigma = \frac{N}{N_e} \times \xi \times \left(\frac{w_e}{w}\right)^{p-1} \times \left(\frac{A}{A_e}\right)^{1-b} \qquad (7.3b)$$

In this expression the first term accounts for the difference in number of mask layers or levels, the second term is the complexity scale factor, which reflects an assessment of how much more or less complicated the new process steps are than the old, the third term is the sensitivity scale factor (or ratio of fault probabilities), with p usually taken to be 3, and the last term is the area scale factor for which a value of $b = 0.5$ is usually adequate for an initial sizing. With this value for λ, the predicted net yield of the new product is

$$Y = e^{-\lambda} = Y_e^{\sigma} \qquad (7.4)$$

For an early sizing, when the design layout is still fluid, such a prediction is sufficient to provide some direction as to whether it is worth pursuing further. It is not much more effort, however, to try to predict the yield based on a layered model.

7.3 LAYERED YIELD MODEL

For specificity, assume that the proposed product is to be built in a silicon bipolar technology. Then the yield losses will be due not only to the photo mask levels, but also to leakage yield losses, which include leakage due to diffusion pipes, emitter to base, and perhaps isolation pockets, which we shall group here as other leakage. Photo layer yield losses are affected by pattern minimum dimensions, but leakage losses are due to point defects. A third source of yield loss is due to miscellaneous defects, such as scratches, foreign material, and other gross defects, which, in general, tend to be larger than most design features. It is convenient to use the notation P_i, $i = 1, 2 \ldots N$ to describe the photo layers; L_i, $i = 1, 2, 3$, to describe leakage due to diffusion pipes, emitter to base, and other mechanisms; and M_i, $i = 1, 2, 3$ to denote scratches, foreign material, and other miscellaneous defects. It does not matter if the proposed product has a different number of photo layers, leakage mechanisms, or miscellaneous defect types than the existing product, since each yield component on the new product will be scaled to the yield component most like it in the existing product.

As part of the ongoing failure analysis of a few wafers from each lot, the number of chips that have failed due to each of the reasons indicated above is recorded and the particular defect-limited yield is computed. As indicated earlier, because it is often not possible to determine which fault occurred first, the sum of the chips failing for different reasons may be more than the total number of chips that have failed. It will be shown later how to correct, at least partially, for the effect of this multiple counting on the yield.

The observed yields of the existing product can then be written as

$$Y_{e,P_i} = 1 - \frac{n_{P_i}}{n_T}, \quad i = 1, 2 \ldots N_e \tag{7.5a}$$

$$Y_{e,L_i} = 1 - \frac{n_{L_i}}{n_T}, \quad i = 1, 2, 3 \tag{7.5b}$$

and

$$Y_{e,M_i} = 1 - \frac{n_{M_i}}{n_T}, \quad i = 1, 2, 3 \tag{7.5c}$$

from which the average number of faults associated with each can be evaluated by taking the negative natural logarithm of each of the expressions above, as was done in Equation (7.2).

The average number of faults for each of these types of faults for the new product is again obtained by scaling, but now the scale factors are not all the same. For the photo faults and miscellaneous faults, the chip active area A may be used as the relevant area, whereas, for the leakage faults, the total emitter area EA is the relevant area. The area scale factors are

$$\alpha_{P_i} = \left(\frac{A}{A_e}\right)^{1-b} \tag{7.6a}$$

$$\alpha_{L_i} = \left(\frac{EA}{EA_e}\right)^{1-b} \tag{7.6b}$$

and

$$\alpha_{M_i} = \left(\frac{A}{A_e}\right)^{1-b} \tag{7.6c}$$

Photo yield loss is caused by defects of finite size, so the photo sensitivity scale factor, which is related to the ratio of fault probabilities can be well approximated by

$$\psi_{P_i} = \left(\frac{w_{e,i}}{w_i}\right)^{p-1} \tag{7.7a}$$

Leakage in bipolar products is usually due to the existence of point defects, such as contaminants of atomic dimension, or dislocations in the silicon lattice, which serve as paths for the leakage of current from one region to another that is meant to be electrically separate. The sensitivity of the device to leakage current determines whether a particular leakage path will result in device failure. Therefore, one way to obtain a leakage sensitivity scale factor is to multiply the number of leakage faults by the ratio of the leakage current threshold $i_{e,t}$ of the existing product to the threshold i_t of the proposed product, raised to an empirically determined power less than unity to account for nonlinearity. Thus,

$$\psi_{L_i} = \lambda_{e,L_i} \times \left(\frac{i_{e,t}}{i_t}\right)^c \tag{7.7b}$$

Miscellaneous defects are usually much larger than the minimum feature dimension, so the miscellaneous sensitivity scale factor can be set equal to unity; that is,

$$\psi_{M_i} = 1 \tag{7.7c}$$

The complexity scale factor is a subjective one, indicating how much more complicated a process step is expected to be for the proposed new product than for the reference product, and a value between 0.90 and 1.10 is usually adequate to account for fabrication differences.

The overall scale factors can then be written as

$$\sigma_{P_i} = \alpha_{P_i} \times \psi_{P_i} \times \xi_{P_i}, \quad i = 1, 2 \ldots N \tag{7.8a}$$

$$\sigma_{L_i} = \alpha_{L_i} \times \psi_{L_i} \times \xi_{L_i}, \quad i = 1, 2 \ldots 3 \tag{7.8b}$$

and

$$\sigma_{M_i} = \alpha_{M_i} \times \psi_{M_i} \times \xi_{M_i}, \quad i = 1, 2 \ldots 3 \tag{7.8c}$$

At each layer, the expected number of photo faults is

$$\lambda_{P_i} = -\sigma_{P_i} \times \ln Y_{e,P_i}, \quad i = 1, 2 \ldots N \tag{7.9a}$$

the expected number of leakage faults is

$$\lambda_{L_i} = -\sigma_{L_i} \times \ln Y_{e,L_i}, \quad i = 1, 2 \ldots 3 \tag{7.9b}$$

and the expected number of miscellaneous faults is

$$\lambda_{M_i} = -\sigma_{M_i} \times \ln Y_{e,M_i}, \quad i = 1, 2 \ldots 3 \tag{7.9c}$$

From this, the respective yields are predicted to be

$$Y_{P_i} = e^{-\lambda_{P_i}}, \quad i = 1, 2 \ldots N \tag{7.10a}$$

$$Y_{L_i} = e^{-\lambda_{L_i}}, \quad i = 1, 2, 3 \tag{7.10b}$$

and

$$Y_{M_i} = e^{-\lambda_{M_i}}, \quad i = 1, 2, 3 \tag{7.10c}$$

The net photo, leakage, miscellaneous yield and overall yields are

$$Y_P = \prod_{i=1}^{N} Y_{P_i}, \quad i = 1, 2 \ldots N \tag{7.11a}$$

$$Y_L = \prod_{i=1}^{3} Y_{L_i}, \quad i = 1, 2, 3 \tag{7.11b}$$

$$Y_M = \prod_{i=1}^{3} Y_{M_i}, \quad i = 1, 2, 3 \tag{7.11c}$$

and

$$Y_{net} = Y_P \times Y_L \times Y_M \tag{7.11d}$$

The advantage of using scaling factors to scale the number of faults of each type inferred from the reference product is that variations in the spatial and size distributions can be accounted for in a manner that allows use of the Poisson yield equation, with its computational convenience, while still providing accurate yield predictions. It is advantageous to use the layered model approach because it not only makes use of more information from the reference product, but also allows each component of the net yield to be computed separately. Thus, when examining the results of these computations, if any yield component appears to be noticeably different from what experience would suggest, assumptions that have been used in its evaluation can be revisited and modified where needed. Furthermore, a layered set of predicted yields provides insight into which are the major factors of yield loss and, together with early data from test chips, can provide timely design and manufacturing guidance before large-scale production commences.

Although, to preserve generality, numerical values have not been used above, the net yield predicted by Equation $(7.11d)$ using a layered model approach may well be different from that of Equation (7.4) using the composite model approach. One way to address the discrepancy is to obtain the ratio of the product of the individual yields of the reference product to the net yield, and apply the same ratio to the new product. This presumes that fatal defects on the new product occur in the same sequence as in the reference product, a presumption that has already been made by selecting the existing product as the reference to which the proposed product is to be scaled.

7.4 ORGANIZATION OF YIELD AND DESIGN DATA

The methods shown in the previous section for predicting the yield of the new product can be used more easily if the available design and yield data and the information that needs to be computed are organized systematically. One way of doing so is shown in Table 7.1, where the available design and yield information about the reference product and the information computed using the equations discussed earlier are listed. Construction of such a table enables all the relevant data, as well as the

Table 7.1

Details of the Existing Reference Product

Layer	Area	Linewidth	Failed Chips	Yield	Faults
P_1	A_{P_1}	w_{P_1}	n_{P_1}	$1 - n_{P_1}/N_T$	$-\ln(1 - n_{P_1}/N_T)$
...					
L_1	A_{L_1}	w_{L_1}	n_{L_1}	$1 - n_{L_1}/N_T$	$-\ln(1 - n_{L_1}/N_T)$
...					
M_1	A_{M_1}	w_{M_1}	n_{M_1}	$1 - n_{M_1}/N_T$	$-\ln(1 - n_{M_1}/N_T)$
...					
Photo	A_P	w	n_P	$1 - n_P/N_T$	$-\ln(1 - n_P/N_T)$
Lkge	A_L	w	n_L	$1 - n_L/N_T$	$-\ln(1 - n_L/N_T)$
Misc	A_M	w	n_M	$1 - n_M/N_T$	$-\ln(1 - n_M/N_T)$
Y_{OA}	A	w	n_{OA}	$1 - n_{OA}/N_T$	$-\ln(1 - n_{OA}/N_T)$

computed values, to be grouped together in one place. Errors in computation can then be readily noticed and corrected.

The yield of each component is unity decreased by the ratio of failing chips to the total number of chips tested. This may result in, for example, the net photo yield being different than the product of the yield of its individual layers, or the overall yield being different than the product of photo, leakage, and miscellaneous yields. The discrepancy is an artifact of the fact that to avoid loss of information, even when a chip is known to have failed due to a fault at a particular layer, faults attributable to subsequent process steps are also counted. However, as the product matures, the number of faults decreases and the discrepancy becomes negligible.

The next step is to construct Table 7.2 for the new product along essentially the same lines as Table 7.1, with a few changes as indicated. Again, the equations developed in the previous section can be used to compute the various terms. In particular, as the scale factors are dimensionless, the appropriate values can be added to obtain the scale factors for photo faults, leakage faults, miscellaneous faults, and all faults. An advantage of a table such as this is that once actual yield data starts being obtained, comparisons can be made with the predicted values, from which more representative values can be determined for the exponent b in the area scale factor, the exponent p in the sensitivity scale factor, and for ξ, the experiential complexity factor. Comparison of predicted values of each defect-limited yield component with the actual yield, if performed and recorded on an ongoing basis for all products being manufactured, provides a valuable experiential base for more accurate yield predictions.

The values of area, linewidth, and the individual scale factors are inserted into the appropriate spaces, and the total scale factor is computed. For the composite

Table 7.2
Details of the Proposed Product

Layer	Area	w	ξ	ψ	α	σ	λ	Y
P_1	A_{P_1}	w_{P_1}	ξ_{P_1}	ψ_{P_1}	α_{P_1}	σ_{P_1}	λ_{P_1}	Y_{P_1}
...								
L_1	A_{L_1}	w_{L_1}	ξ_{L_1}	ψ_{L_1}	α_{L_1}	σ_{L_1}	λ_{L_1}	Y_{L_1}
...								
M_1	A_{M_1}	...	ξ_{M_1}	ψ_{M_1}	α_{M_1}	σ_{M_1}	λ_{M_1}	Y_{M_1}
...								
Photo	A_P	w_P	ξ_P	ψ_P	α_P	σ_P	λ_P	Y_P
Lkge	A_L	w_L	ξ_L	ψ_L	α_L	σ_L	λ_L	Y_L
Misc	A_M	...	ξ_M	ψ_M	α_M	σ_M	λ_M	Y_M
Y_{OA}	A	w	ξ_{OA}	ψ_{OA}	α_{OA}	σ_{OA}	λ_{OA}	Y_{OA}

photo and composite leakage, an average linewidth may be used. For the miscellaneous case, the linewidth does not enter the calculation and so is left blank. Having obtained σ and selecting the appropriate value of faults from Table 7.1, the expected number of faults is computed and the Poisson equation is used to predict the individual layer yields.

As in the case of Table 7.1, the overall yield shown in Table 7.2 may not agree with the product of the yield of each of the individual layers. The way to address this discrepancy has been indicated earlier. However, over time the product matures and the discrepancy between the overall composite yield and the product of the individual layer yields rapidly becomes much less.

7.5 YIELD LEARNING

A complete analysis requires yield prediction not only at the start of production but over the expected manufacturing life of the product. Initially, the yield will tend to be low, but as experience is gained, many of the defects that lower the yield will be reduced. Thus, predicting later yield essentially involves predicting the rate at which the incidence of defects is expected to decrease. There are many theoretical learning curves, but, in practice, the rate at which yield increases or defects are reduced depends on two factors: the care with which the fabrication is carried out and the resources available to purchase new tools likely to generate fewer defects. The second factor depends on business considerations that are not always readily amenable to quantification, but the former can be readily quantified by tracking the yield of each product at each layer and plotting either the number of faults of each type per chip, or the percentage reduction in this number, over some convenient time period. Three months (i.e., one quarter) is a convenient time period that is often used to average over, even though, for maximum benefit, tracking should be performed on a regular, ongoing basis and reported weekly.

If for a particular fabrication facility it has been found that faults of type i tend to be reduced by a factor L_{iq} each quarter, where q denotes the number of quarters after start of normal production, then (if the Poisson yield equation is used) the ith defect-limited yield in the qth quarter can be expressed in terms of the yield at start of normal production by the equation

$$Y_{iq} = e^{-\lambda_{i0}/L_{iq}} = Y_{i0}^{1/L_{iq}} \qquad (7.12)$$

Because the rate of defect reduction usually differs from one defect type to another, it is preferable to use a layered yield model to predict yield than a composite one. But it is not necessary to be restricted to the use of the Poisson yield equation. The method discussed in this chapter can be used with any yield equation as long as it is used consistently. However, inferring the number of faults of each type per chip from the observed yield of the reference product, or computing the yield of the new product from the scaled number of faults of each type, requires more effort, and it is not clear that it will lead to improved accuracy in the predicted yield. The Poisson yield equation is computationally simple and its terms admit of simple physical interpretation. The objection that it predicts yields lower than is later observed has not been found to be valid, at least for the cases reported [1–3], if the defect density is scaled as shown to accommodate both the spatial variation of the defect density, as well as the variation of defect sizes. In particular, the use of a layered Poisson yield model enables each of the terms affecting the predicted yield to be clearly delineated in a manner that allows them to be modified objectively as actual data become available, thus permitting the accuracy of the predictions to improve.

7.6 REFERENCES

[1]   A.V. Ferris-Prabhu, "Forecasting Semiconductor Yield," *IEEE International Conference on Computers, Systems, and Signal Processing*, Indian Institute of Science, Bangalore, India, December 9–12, 1984, Paper No. R46.13, Vol. 3, pp. 1491–1497.

[2]   A.V. Ferris-Prabhu and J.A. Prabhu, "A Practical Method of Yield Estimation," *Proceedings of the Second International Workshop on VLSI Design*, Bangalore, India, December 15–18, 1988, R. Apte, V.D. Agarwal, and A. Prabhakar, eds., pp. 71–79.

[3]   A.V. Ferris-Prabhu and M. Retersdorf, "The Effect on Yield of Clustering and Radial Variations in Defect Density," in *Defect and Fault Tolerance in VLSI Systems*, Vol. 2, C.H. Stapper, ed., Plenum Press, 1990, pp. 63–73.

Chapter 8
Yield With Redundancy

8.1 INTRODUCTION

This chapter builds on the material covered in the previous chapter to explain how to calculate the yield of products such as memory array products that have on-chip redundancy.

A memory chip usually has a large number of identical array segments that serve as storage, and a small amount of decode and peripheral circuitry common to the entire chip. If either peripheral circuitry or decode circuitry fails, the chip fails. If an array segment fails, depending on how the chip is organized, the chip may or may not fail. If there are exactly as many array segments as are needed for a word, then if an array segment fails, the chip fails. However, the cost is small, in terms of extra silicon or extra effort, to fabricate the chip with one or more extra array segments, depending on how the chip is organized. If all segments are good, only those that are needed are accessed. If one or more segments fail, then the spares are switched on and the chip is still functional and does not need to be discarded. For the purpose of this chapter, we will consider chips that are fabricated with N array segments, of which only $N - 1$ are needed for the chip to function. It is customary for the yield of chips with all array segments good to be referred to as the all-good yield Y_{AG}, and with an adequate number of segments good as the partial-good yield Y_{PG}. If, as in this case, a partially good chip is equivalent functionally to an all-good chip, the equivalent yield Y_{EQ} is the sum of the all-good and partial-good yields. For other chip organizations, two or more partially good chips may be needed to be equivalent to one all-good chip with appropriate change in the definition of the equivalent yield.

It is convenient to discuss the yield in terms of the array yield $Y_{ARR,N}$ of all N segments, or $Y_{ARR,N-1}$ of $N - 1$ segments, and the peripheral yield Y_{PER}. For a chip to be good, the peripheral circuitry must be good, and either all array segments or all but one array segments need to be good. Therefore, predicting the yield becomes a question of calculating the probability that the entire peripheral circuitry will be good and the probability that N array segments or any $N - 1$ array segments are good. These probabilities depend not only on the design of the product (i.e., its pattern geometry, including the fraction of the chip that is array and the fraction that is peripheral), which is known, but also on the defect density, which is not known but needs to be scaled from an existing product or test structures. Making use of the material presented in the previous chapter, this chapter will give expressions for the yield both in terms of the scaled numbers of faults and in terms of the scaled yield of the reference product. A significant advantage of the Poisson yield equation is that it permits this scaling to be done very simply. To simplify the treatment, we shall first consider a composite model where only the chip as a whole is considered, and then we shall consider what might be called, for lack of a better name, a partially layered model in which only the photo-limited yield, leakage-limited yield, and the miscellaneous defects-limited yield, each taken as a whole, are used in the calculation.

8.2 COMPOSITE YIELD MODEL

The all-good yield can be formally written as

$$Y_{AG} = Y_{ARR,N} \times Y_{PER} \tag{8.1}$$

and the partial-good yield as

$$Y_{PG} = Y_{ARR,N-1} \times Y_{PER} \tag{8.2}$$

In terms of the yield Y_a of a single array segment, the yield of all N array segments can be written as

$$Y_{ARR,N} = Y_a^N \tag{8.3}$$

and, making use of the binomial expansion, the yield of $N - 1$ array segments can be written as

$$Y_{ARR,N-1} = N Y_a^{N-1}(1 - Y_a) \tag{8.4}$$

where $(1 - Y_a)$ is the probability of one segment being bad.

It is convenient to rewrite Equation (8.4) as

$$Y_{ARR,N-1} = N Y_{ARR,N}(Y_{ARR,N}^{-1/N} - 1) \tag{8.5}$$

Inserting Equation (8.5) into Equation (8.2) and making use of Equation (8.1), the partial good yield is given by

$$Y_{PG} = Y_{AG} N(Y_{ARR,N}^{-1/N} - 1) \tag{8.6}$$

The partial-good yield can never be negative because no yield term can exceed unity and the reciprocal of a number less than unity is greater than one. So the term in parentheses is usually positive but cannot be less than zero.

The total or equivalent yield is then

$$Y_{EQ} = Y_{AG}[1 + N(Y_{ARR,N}^{-1/N} - 1)] \tag{8.7}$$

In this representation the contribution of the partial-good yield is clearly seen.

If the array portion of the chip is a fraction f of the all-good chip, the equivalent yield can also be written as

$$Y_{EQ} = Y_{AG}[1 + N(Y_{AG}^{-f/N} - 1)] = e^{-\lambda_{AG}}[1 + N(e^{+f\lambda_{AG}/N} - 1)] \tag{8.8}$$

where

$$\lambda_{AG} = \sigma\lambda_e \tag{8.9a}$$

$$\lambda_e = -\ln Y_e \tag{8.9b}$$

and σ is the composite model scale factor introduced and discussed in the previous chapter. In terms of Y_e, the yield of the existing product being used as the reference database to which the yield of the new product is being scaled, the equivalent yield can be written as

$$Y_{EQ} = Y_e^{\sigma}[1 + N(Y_e^{-f\sigma/N} - 1)] \tag{8.10}$$

It should be noted that the convenient representation of Equation (8.10) is applicable only if the Poisson yield equation is used. However, the methods of this section can be used no matter which yield equation is used, though with equations other than the Poisson, the calculation of the average number of faults per chip becomes lengthy and the final expression for equivalent yield is cumbersome and not as readily amenable to an examination of its constituent terms.

8.3 PARTIALLY LAYERED YIELD MODEL

More detail can be obtained by using a model in which the contributions of the photo-limited, the leakage-limited, and the miscellaneous defect-limited yields are taken into account separately instead of being lumped together. Doing so allows the all-good yield to be written as

$$Y_{AG} = Y_P \times Y_L \times Y_M \tag{8.11}$$

If f_P is the fraction of the chip in the array segment that is susceptible to photo faults, f_L the fraction of the chip in the array segment that is susceptible to leakage faults, and f_M the fraction of the chip in the array segment that is susceptible to miscellaneous faults, the yield of all N array segments can be written as

$$Y_{\text{ARR},N} = Y_P^{f_P} \times Y_L^{f_L} \times Y_M^{f_M} \tag{8.12}$$

Using Equation (8.12), the equivalent yield given by Equation (8.7) can be written as

$$Y_{\text{EQ}} = Y_P Y_L Y_M [1 + N(Y_P^{-f_P/N} Y_L^{-f_L/N} Y_M^{-f_M/N} - 1)] \tag{8.13}$$

where

$$Y_P = e^{-\lambda_P} = e^{-\sigma_P \lambda_{e,P}} = Y_{e,P}^{\sigma_P} \tag{8.14a}$$

$$Y_L = e^{-\lambda_L} = e^{-\sigma_L \lambda_{e,L}} = Y_{e,L}^{\sigma_L} \tag{8.14b}$$

and

$$Y_M = e^{-\lambda_M} = e^{-\sigma_M \lambda_{e,M}} = Y_{e,M}^{\sigma_M} \tag{8.14c}$$

In these equations, the left-hand side is the predicted yield components of the proposed new product, scaled to the observed yield of an existing product to which it is being scaled, making use of the scale factors discussed in the previous chapter.

The equivalent yield can also be written in terms of a fully layered model that includes each of the photo mask levels, each of the leakage components, and each of the miscellaneous defect components. The expression will be lengthier, but will provide more detail. Such detail is advantageous, particularly if yield estimates are made early in the design phase, because each component of the yield can be related to parameters under the control of the designer, such as the areas sensitive to photo, leakage, or miscellaneous defects; the fraction of the chip that is array; the fraction used for peripheral and decode circuitry; and the minimum design dimensions. By carefully analyzing the yield terms and the design parameters that enter into them, design changes that will enhance yield can be effected before fabrication starts. Furthermore, after production starts, a careful record of each of the yield components will not only strengthen the database for future products, but will also provide much information about experiential factors, such as the exponent b in the area scaling factor. Although yield prediction has a sound theoretical basis, the values of many of the terms that enter into it are still not readily available from theoretical considerations and need to be obtained from the experience and insight gained by analyzing actual data.

8.4 REFERENCE

[1] A.V. Ferris-Prabhu, "Reliability Enhancement with Error Correction," *Computer Design*, July 1979, pp. 137 ff.

Chapter 9
A Yield Comparison

9.1 INTRODUCTION

This chapter shows a generalized yield equation that encompasses the negative binomial and the regular Poisson equations, discusses assumptions made, but not usually explicitly stated, in most yield equations, explains the difficulties of verifying which equations have better predictive capability and discusses an attempt to fit yield data by each of two commonly used yield equations.

9.2 A GENERAL YIELD EQUATION

It was mentioned in Chapter 5 that a feature common to many problems is that of dispersion: dispersion of plant species, of human characteristics, of galaxies, of semiconductor defects, and, interestingly enough, of retail stores in cities. A study of dispersion is beyond the scope of this book, but interested readers may refer to an elegant statistical analysis of spatial dispersion in the excellent text by A. Rogers [1], who shows that the form of the expression predicting the occurrence of an event in a given region depends on whether the probability of occurrence of an additional event is related to or independent of the prior occurrence of other events in that region. From Rogers' treatment it can be shown in particular that the probability of no event occurring is succinctly contained in the equation

$$p(0,\lambda) = (1 + \beta\lambda)^{-1/\beta} \tag{9.1}$$

where in conformity with previous notation, λ, the average number of events in a region may be interpreted as the average number of fatal defects per chip, $p(0,\lambda)$ as the yield, and β as the ratio of the variance of the defect probability density function to the square of its average value. In a more physical sense, β may be interpreted as a coefficient coupling the probability of occurrence of events, with $\beta = 0$ implying no coupling and $|\beta| = 1$ implying maximum coupling, a negative value of β meaning the coupling is repulsive and β positive implying that it is attractive. Specifically, with $\beta = 0$, Equation (9.1) reduces to the Poisson equation

$$p(0,\lambda) = e^{-\lambda} \tag{9.2}$$

When $\beta = -1$, Equation (9.1) becomes

$$p(0,\lambda) = 1 - \lambda \tag{9.3}$$

which is similar to the expression derived from application of the Fermi-Dirac statistics, according to which at most one particle can occupy a state, an expression that is not suitable as a yield equation.

When $\beta = +1$, Equation (9.1) becomes

$$p(0,\lambda) = \frac{1}{1 + \lambda} \tag{9.4}$$

which is similar to the expression derived from application of the Bose-Einstein statistics, according to which there is no limit on the number of particles that can occupy a state. Thus, the interpretation of β as a coupling coefficient is reasonable, and the generality of Equation (9.1) is seen.

However, the defects that occur in semiconductor device fabrication are not Bose particles, though more than one may occur on the same chip. Thus, it is appropriate to use a positive value less than unity for β. A value of 0.5 has been reported [2] to provide accurate yield predictions.

Interpreting Equation (9.1) as a yield equation and inverting it shows that the number of faults per chip can be obtained from the known yield by using the equations

$$\lambda = (Y^{-\beta} - 1)/\beta, \quad \beta \neq 0 \tag{9.5a}$$

and

$$\lambda = -\ln Y, \quad \beta = 0 \tag{9.5b}$$

The effect of the coupling coefficient β can then be examined by plotting the ratio of Equation (9.5a) to Equation (9.5b) (i.e., the relative number of faults) for different values of β or of yield. This is done in Figure 9.1, which shows the number of faults per chip for positive values of the coupling coefficient β normalized to the number of faults per chip when $\beta = 0$. Note that for a given value of yield, there is a higher value for the number of faults per chip with higher values of β.

Alternatively, it is possible to plot the yield versus the number of faults per chip with different values of β, as shown in Figure 9.2.

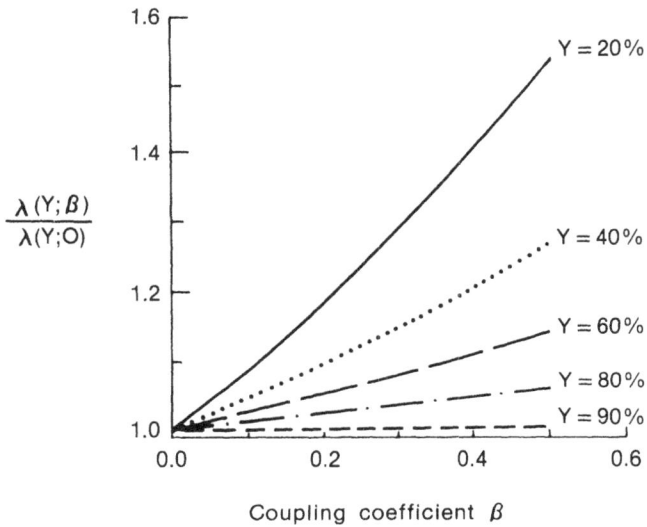

$$\frac{\lambda(Y;\beta)}{\lambda(Y;O)}$$

Figure 9.1 Relative number of faults per chip *vs.* coupling coefficient β.

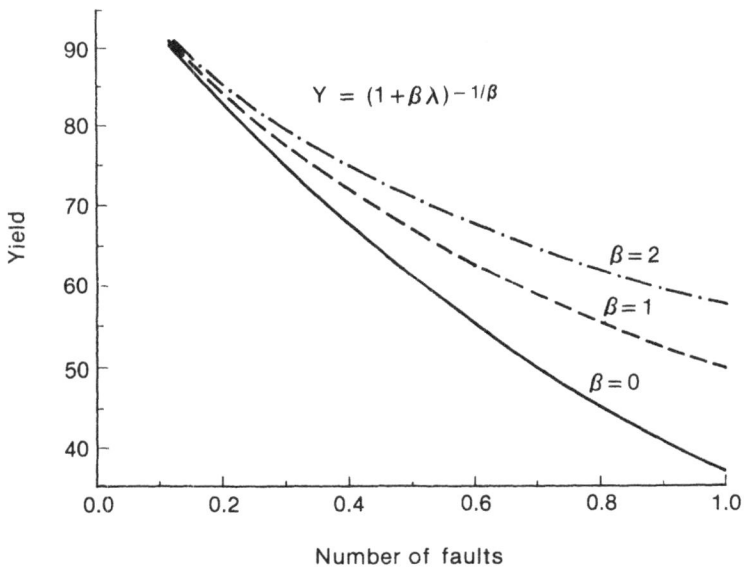

$$Y = (1+\beta\lambda)^{-1/\beta}$$

Figure 9.2 Yield *vs.* faults per chip for different values of the coupling coefficient β.

Figures 9.1 and 9.2 show the effect of the coupling coefficient β. For a given number of faults per chip, a larger value of β gives a higher yield, and, conversely, for a given yield, a larger value of β gives a larger average number of faults per chip. These observations reinforce the interpretation of β as a coefficient coupling the occurrence of fatal defects. Large values of β imply a stronger coupling between faults, which in turn implies that the total number of faults on a wafer are distributed over fewer chips, resulting in higher yield. It is the presence of this adjustable parameter β that gives the negative binomial yield equation a greater flexibility than other yield equations, but at the cost of lower computational efficiency.

9.3 ASSUMPTIONS AND DIFFICULTIES COMMON TO YIELD EQUATIONS

These results suggest that it may be useful to examine different yield equations for the assumptions they contain. Such an analysis [3] shows that two major assumptions are common to all yield equations. Assumptions about the spatial distribution of potentially fatal defects determine the form of the defect probability density function $f(D)$ and thus the form of the yield equation, as shown in Chapters 4 and 5. Also, as shown in Chapter 3, assumptions about the size distribution of potentially fatal defects determine the magnitude of the fault probability and thus of the average number of fatal defects used in the argument of the yield equation.

The difficulty is that neither the spatial nor the size distribution of potentially fatal defects is as yet amenable to unambiguous direct determination on actual product wafers. Various methods have been proposed in the literature for the determination of the defect density and of the sizes of defects on test chips that have been designed specially to "capture" defects, "capture" being determined by whether or not a particular circuit passes test. In other words, the loss of yield is an indication of the presence of a defect. However, it is not clear [4] that it is possible from yield data alone to infer both the spatial distribution of all potentially fatal defects and their size distribution [5]. Development of unambiguous methods of determining these quantities is a fruitful area for further research.

A third and major difficulty is that design and yield data are proprietary, so information about predicted and actual yield and about the results of the analyses of failing chips is not available in the literature in the detail needed either to confirm a model or to determine where it is in error. In many cases there is no doubt that a particular yield equation does not describe actual data, but the predictive capabilities of a yield equation are less clear.

9.4 A COMPARATIVE YIELD ANALYSIS

In this section we will use two yield equations, the Poisson yield equation with no clustering ($\beta = 0$) and the negative binomial yield equation ($0 \leq \beta \leq 1$), to "predict"

the yield of a ten-array segment bipolar static random access memory (SRAM) product for which yield data exist [6]. This example has been chosen because this chip
has ten identical array segments that can be treated as pseudochips of one, two, or
more array segments each, the yield of each of which is known. The yield of the
one-array segment pseudochip is used as the reference from which to "predict" the
yield of the actual chip, whose known yield can serve to verify the accuracy of the
"prediction."

In this analysis [3], the random defect-limited yield Y_r is defined as the ratio
of the observed yield Y_{obs} to the gross cluster or miscellaneous defect-limited yield
Y_0. In general, the Y_0 term is obtained by extrapolating to zero area, the yield of
successively smaller grids on a wafer yield map. For the ten-array segment SRAM
chip being discussed, extrapolating the curve shown in Figure 9.3 to zero area shows
that Y_0 is almost unity, suggesting that for this product, which happens to be a mature
one, the random defect-limited Y_r is essentially the same as the observed yield Y_{obs}.

Because each size pseudochip is identical to every other size pseudochip in all
respects except that of area, the complexity and sensitivity scale factors are unity.
The area scaling factor

$$(A/A_1)^{1-b}$$

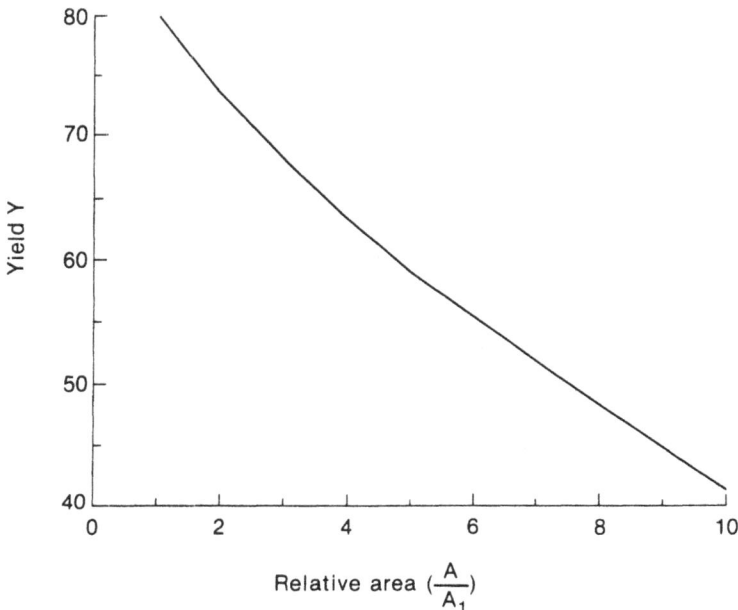

Figure 9.3 Yield of pseudochips of different sizes *vs.* their relative area.

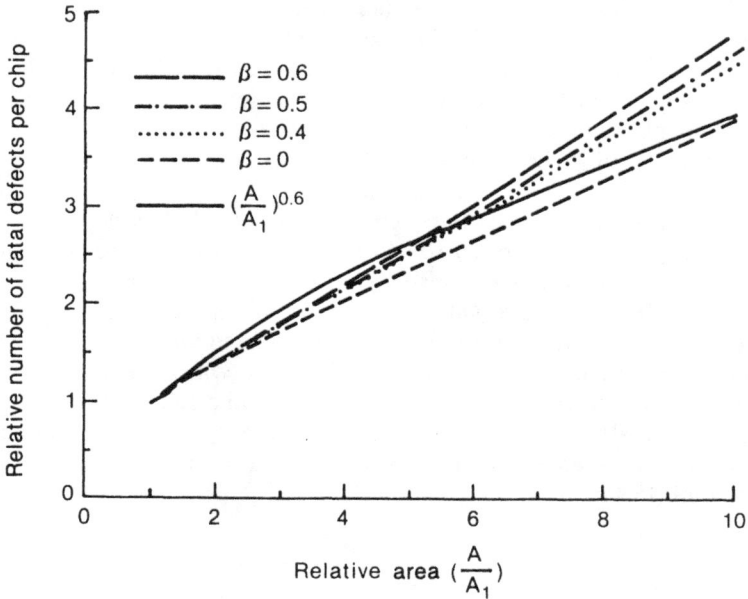

Figure 9.4 Relative number of faults per pseudochip *vs.* relative area. The area scale factor is also
plotted.

is obtained by reading off from Figure 9.3 the values of yield for different size
pseudochips, computing the average number of faults per pseudochip relative to the
number of faults per single array segment pseudochip, and finding the exponent b
in the area scale factor that provides the best fit for different values of β. Figure 9.4
shows that the exponent $b = 0.4$ in the area scale factor agrees fairly well with the
relative number of faults per chip for four values of β.

This suggests that the yield of multiple array segment pseudochips of area A
can be predicted from the observed yield of a single array segment pseudochip of
area A_1 by using the generalized yield equation

$$Y(A) = [1 + \beta\lambda(A)]^{-1/\beta} \tag{9.6a}$$

where

$$\lambda(A) = \lambda(A_1) \times (A/A_1)^{0.6} \tag{9.6b}$$

and

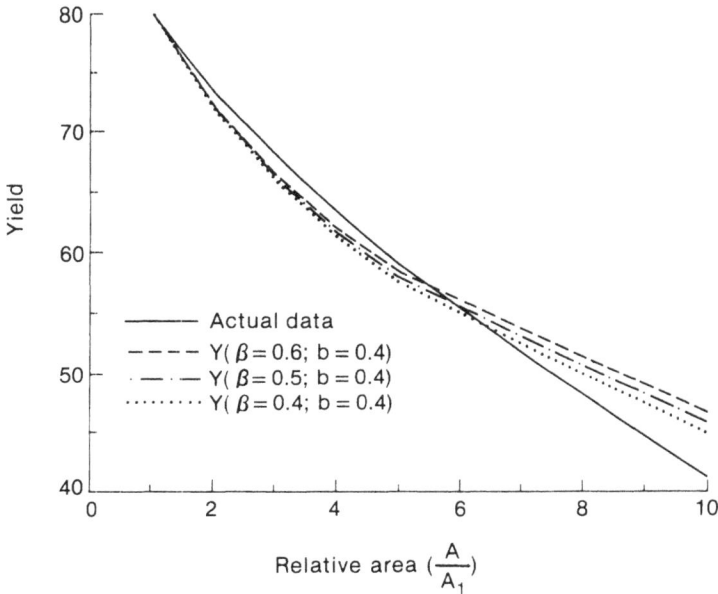

Figure 9.5 Actual and predicted yields *vs.* relative area with $b = 0.4$.

$$\lambda(A_1) = [Y(A_1)^{-\beta} - 1]/\beta, \quad \beta \neq 0 \tag{9.7a}$$

and

$$\lambda(A_1) = -\ln Y(A_1), \quad \beta = 0 \tag{9.7b}$$

With $b = 0.4$, the predictions of the generalized yield equation with four values of the coupling coefficient β compare favorably with the actual yield, as shown in Figure 9.5.

Decreasing the value of b in the area scale factor depresses the predicted yield, whereas increasing β increases the predicted yield slightly. By adjusting the values of β and b, the average number of fatal defects per single array segment pseudochip inferred from its yield can be altered, and these same values can then be used to predict the yield of larger array segments. The predictions shown in Figure 9.6 for the negative binomial yield equation ($\beta = 0.4$) with $b = 0.4$ and for the Poisson equation ($\beta = 0$) with $b = 0.45$ both agree well with the actual yield.

It is clear that more than one choice of coupling coefficient and of value for b in the area scale factor predicts yields in quite reasonable agreement with the actual yield. In this case, where the actual yield is known already, the parameters b and

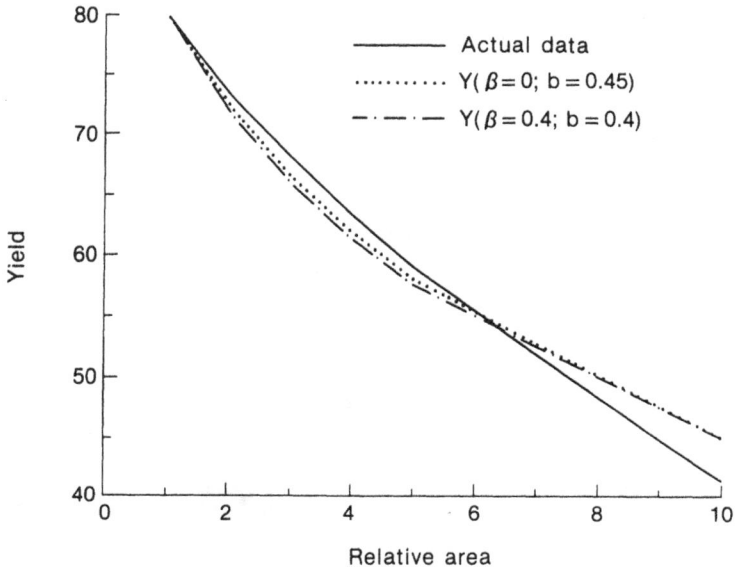

Figure 9.6 Actual and predicted yields. For the negative binomial model ($\beta \neq 0$) with $b = 0.4$ and for the Poisson model ($\beta = 0$) with $b = 0.45$.

β, or even the term Y_0, can be "tuned" to fit the data. However, the objective of this analysis is not to show which model can better *describe* known results, but to show that the *predictive* capability depends not only on the choice of model, but also on the manner in which the average number of fatal defects per chip of existing product is scaled to the new product. In the example discussed, only the area scale factor differed from unity, but in a more realistic prediction, where the new product is likely to have different minimum design features and may well be built in a different process, the process complexity and defect sensitivity scale factors will need to be used as well. A reason why the Poisson model has been reported to predict lower yields than have been observed is that its users may not have adequately scaled the number of faults per chip inferred from an existing product. As has been shown here, such a scaling enhances the predictive capability of the yield model used.

9.5 DISCUSSION

To predict yield for a semiconductor product, essentially two pieces of information are needed: a yield equation and the average number of fatal defects per chip to use as the argument of the yield equation. The form of the yield equation depends on how the defects are dispersed over the wafer and reflects the assumptions made about the nature of the interaction between defects. The interaction may be inhibitory (Fermi-Dirac statistics), which is not likely, or attractive (Bose-Einstein statistics), which is more likely, or there may be no interaction at all (Maxwell-Boltzmann statistics or Poisson-like) for a mature product. The average number of fatal defects per chip of new product needs to be obtained from the yield of existing product or test structures, and necessarily reflects assumptions about both the size distribution and the spatial dispersion of defects on that product or test structure.

For the independent, sequential process steps typical of semiconductor device fabrication, in the absence of any physical model to the contrary, there is no reason to assume that there is interaction between defects. This leads to the Poisson model. The Poisson model as customarily used, infers the defect density from the known yield of existing product, and uses this value even when the new products are larger than the existing product. The yield thus predicted has been widely reported to be pessimistic. This discrepancy, together with the observation that defects frequently occur in clusters, particularly near the periphery of the wafer, has led to the development of cluster models, the predictions of which have been reported to agree better with later observed results. However, as has been shown, it is not only the yield equation that affects the predicted value of yield, but also the average value of fatal defects per chip used in the yield equation. This average value is not independently known ahead of time; it needs to be scaled from the value *inferred* from the yield of existing product, with the scale factor reflecting differences in process complexity and chip area, as well as defect size and spatial distributions, between new and existing product.

Thus, the accuracy of yield predictions depends just as much on the accuracy of the assumed average number of fatal defects per chip as on the choice of yield model. For the example investigated, when appropriate scaling rules are used, the predictions of the yield equation that assume no coupling between defects are no less accurate than the predictions of the model that assume that the defects are coupled, suggesting that in such cases it may be desirable to use the yield equation that is computationally more efficient.

9.6 REFERENCES

[1]   A. Rogers, *Statistical Analysis of Spatial Dispersion*, Pion Ltd., London, 1974.

[2]   C.H. Stapper, "Fact and Fiction in Yield Modeling," *Microelectronics Journal*, Vol. 20, No. 1/2, Spring 1989, pp. 129–151.

[3]   A.V. Ferris-Prabhu, "On the Assumptions Common to All Yield Models," *IEEE Transactions on Computer Aided Design of Integrated Circuits and Systems*, Vol. 11, No. 8, August 1992.

[4]   A.V. Ferris-Prabhu, "Role of Defect Size Distribution in Yield Modeling," *IEEE Transactions on Electron Devices*, Vol. ED-32, No. 9, September 1985, pp. 1727–1736.

[5]   C.H. Stapper, "Modeling of Integrated Circuit Defect Densities," *IBM Journal of Research and Development*, Vol. 27, No. 6, November 1983, pp. 549–557.

[6]   A.V. Ferris-Prabhu and M. Retersdorf, "The Effect on Yield of Clustering and Radial Variations in the Defect Density," in *Defect and Fault Tolerance in VLSI Systems*, Vol. 2, C.H. Stapper, ed., Plenum Press, 1990.

Chapter 10
Productivity

10.1 INTRODUCTION

This text has presented yield equations and has shown how to predict yield. However, productivity, of which yield is an important, if not the most important, component, is the quantity that determines the success or otherwise of a product. The first chapter defined productivity as the number of packaged chips available per wafer start. In this chapter we will use a slightly different definition which is more useful at the wafer level (i.e., the number of circuits that are available per wafer that has completed fabrication). We will show the components of productivity so defined, present expressions to evaluate them, and provide a set of prescriptions that can give quantitative values for use in developing product strategy.

10.2 PRODUCTIVITY

The productivity at wafer level can be defined [1] as the number of circuits available per wafer that has completed fabrication; that is,

$$P = n \times N \times Y \tag{10.1}$$

where n is the number of circuits per chip, N is the number of chips per wafer, and Y is the fraction of them that are usable (i.e., the yield).

The number of circuits per chip is determined by design parameters such as the chip size s, the minimum design dimension w, the number of levels of vertical integration m, and the design system d, where $d = 1$ is used to denote a gate array design, $d = 2$ to denote a standard cell design, and $d = 3$ to denote a microprocessor design, each of which is denser than the previous one. The number of chips per wafer is a purely geometrical quantity determined by the size s of the chip plus the width k of the kerf and pad cage, and R_e, the effective radius of the wafer. To minimize chip edge damage due to wafer handling and tool placement, chips are fabricated on the wafer such that they lie entirely inside a containment circle of radius

---

R_e, which is about 1 mm less than the actual wafer radius R. The yield Y is determined by the same parameters that determine the number of circuits per chip, plus the defect density. For any given product, all parameters are invariant once the design is fixed, except for the defect density, which is not a design parameter and which decreases with time as experience is gained in the fabrication of a product.

A qualitative representation of the effect of the design parameters on the components of the productivity is given in Figure 10.1 and discussed in detail elsewhere [2].

In the following sections, expressions will be given for the number of circuits per chip, the number of chips per wafer, and the yield, in terms of the same quantities for a product that is already being manufactured. There are two reasons for this scaling approach. One is that some parameters are not known for a new product and have to be estimated from their values in a product already in manufacture. The second is that scaled values do not reveal sensitive information, whereas data that give the actual values of semiconductor design parameters, defect densities, and yields are highly proprietary and are not generally available. Nevertheless, the accuracy and self-consistency of the expressions that will be derived can be checked by letting the parameters take on the value of the known product, in which case the computed terms should agree with the experimentally determined values. For brevity, only the final expressions will be given, and the reader may refer to recent articles [1, 2] for more detail.

For the purpose of this chapter, the reference product used will be a square gate array chip ($d = 1$) of edge s_0 mm and kerf width k_0 mm with two levels of metallization ($m = 2$) and minimum design dimension w_0 μm, that has ν_0 circuits

	N	n	Y	P = NnY
R ↑	↑	–	–	↑
d ↑	–	↑	↓	?
m ↑	–	↑	↓	?
w ↓	–	↑	↓	?
s ↑	↓	↑	↓	?
K ↓	↑	–	–	↑
D ↓	–	–	↑	↑

Figure 10.1 The effect of design parameters on the productivity and its components.

per mm^2 and yield Y_0. The number of such chips that can be fabricated on a wafer of effective radius R_e is defined as N_0.

10.3 CIRCUITS PER CHIP

Referenced to this product, the number of circuits per chip of edge s being designed in design system d with minimum design dimension w and m levels of vertical integration is [1]

$$n_{dmw}(s) = n_0 \left(\frac{s}{s_0}\right)^2 \Gamma_{dmw} \tag{10.2}$$

where

$$n_0 = n_{12w_0}(s_0) = \nu_0 s_0^2 \tag{10.3}$$

The design parameters scale factor is

$$\Gamma_{dmw} = f_d g_m h_w \tag{10.4}$$

where the design system scale factors f_d are approximately

$$f_1 = 1.00, \quad f_2 = 1.13, \quad f_3 = 1.6 \tag{10.5a}$$

the levels of integration or metal levels scale factor g_m is

$$g_m = (\sqrt{2})^{m-2} \tag{10.5b}$$

and the minimum design dimension or ground rules scale factor h_w is

$$h_w = \left(\frac{w_0}{w}\right)^2 \tag{10.5c}$$

With these expressions, the number of circuits per chip can be approximated quite accurately, although as tools and design systems change, the scale factors may need to be modified to reflect these changes.

Table 10.1 shows the number of circuits per chip for several proposed products relative to that on an existing double metal level gate array chip of the same size and with the same ground rules.

Table 10.1
Relative Number of Circuits per Chip

n_{dmw}	$d = 1$	$d = 2$	$d = 3$
$m = 2$	1.00	1.13	1.60
$m = 3$	1.41	1.64	2.26
$m = 4$	2.00	2.26	3.20

10.4 CHIPS per WAFER

The number of chips per wafer is a purely geometrical quantity given by [3]

$$N(L,W) = \frac{\pi R_e^2}{LW} e^{-L/R_e} \tag{10.6}$$

where L and W are the length and width of the chip (in both cases including kerf plus pad cage) on a wafer of effective radius R_e.

Using this prescription, the accuracy of which is shown in Figure 10.2, the number of square chips of active dimension s plus kerf and pad cage k that can fit on a wafer of effective R_e can be expressed in terms of N_0, the number of chips with dimension $s_0 + k_0$ that fit on a similar size wafer; that is,

$$N(s,k) = N_0 \left(\frac{s_0}{s}\right)^2 \times f(s,k,s_0,k_0) \tag{10.7a}$$

where

$$f(s,k,s_0,k_0) = \exp\left[-\frac{(s + k) - (s_0 + k_0)}{R_e}\right] \times \left(\frac{1 + k_0/s_0}{1 + k/s}\right)^2 \tag{10.7b}$$

The number of potentially available circuits per wafer in terms of the reference product is

$$n_{dmw}(s)N(s) = n_0 N_0 \times \Gamma_{dmw} \times f(s,k,s_0,k_0) \tag{10.8}$$

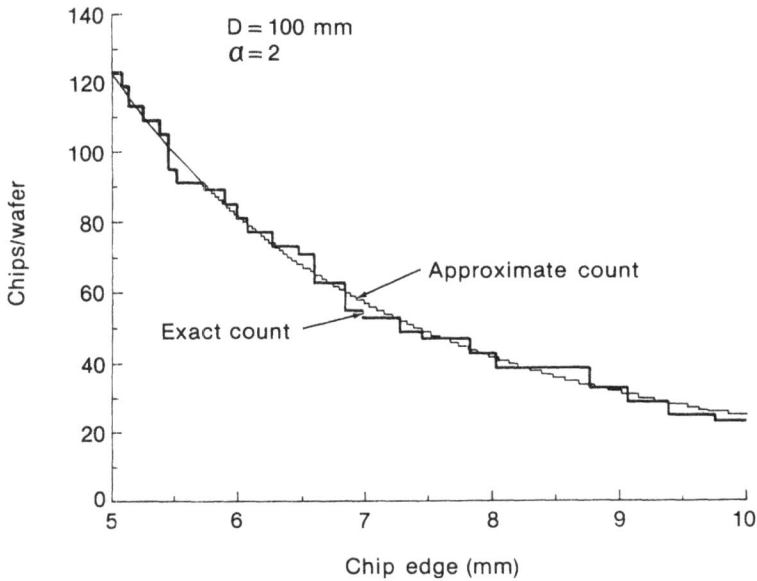

Figure 10.2 Exact and approximate chip count. The chips have a length-to-width ratio of 2 and are fabricated on a wafer 100 mm in diameter.

10.5 YIELD

Using the Poisson yield equation, the yield of the proposed product is predicted to be

$$Y_{dmw}(s) = e^{-\lambda_{dmw}} \tag{10.9}$$

The average number of faults per chip of the proposed design is scaled to the average number of faults per chip of the reference product; that is,

$$\lambda_{dmw}(s) = \sigma\lambda_{12w_0} \tag{10.10}$$

where, as shown in an earlier chapter,

$$\sigma = \xi \times \psi \times \left(\frac{s}{s_0}\right)^{2(1-b)} \tag{10.11}$$

Identifying the complexity scale factor ξ with the product of the design system scale factor f_d and the metal levels scale factor g_m, and the sensitivity scale factor ψ with the ground rules scale factor h_w, shows that the overall scale factor σ can be rewritten as

$$\sigma = \alpha\Gamma_{dmw} \tag{10.12a}$$

where the area scale factor α is given by

$$\alpha = \left(\frac{s}{s_0}\right)^{2(1-b)} \tag{10.12b}$$

Equation (10.10) can then be rewritten as

$$\lambda_{dmw,0}(s) = \alpha\Gamma_{dmw} \times \lambda_{12w_0} \tag{10.13}$$

and the yield of the proposed product as

$$Y_{dmw,0}(s) = [Y_{12w_0}(s_0)]^{\alpha\Gamma_{dmw}} \tag{10.14}$$

where the subscript 0 refers to the yield prior to start of defect reduction. If it is expected that the number of faults will be reduced by half every τ quarters, then after q quarters the average number of faults per chip is given by

$$\lambda_{dmw,q}(s) = \alpha\Gamma_{dmw} \times \lambda_{12w_0} \times 2^{-q/\tau} \tag{10.15}$$

and

$$Y_{dmw,q}(s) = [Y_{12w_0}(s_0)]^{\alpha\Gamma_{dmw}2^{-q/\tau}} \tag{10.16}$$

where the last term in Equation (10.15) and in the exponent of Equation (10.16) shows the effect of yield learning.

The expected yield of several proposed products with the same chip size and ground rules as an existing double level metal gate array chip with 25% yield is shown in Table 10.2.

Table 10.2
Predicted Yield When Yield of Reference Product Is 25%

$Y_{dmw}\%$	$d = 1$	$d = 2$	$d = 3$
$m = 2$	25	21	10.8
$m = 3$	14	10.8	4.4
$m = 4$	6.25	4.4	1.18

10.6 RELATIVE PRODUCTIVITY

The productivity of the proposed product q quarters after start of normal production is therefore

$$P_{dmw,q}(s) = n_{dmw} \times N(s,k) \times Y_{dmw,q}(s) \qquad (10.17)$$

where the terms on the right-hand side have been defined in Equations (10.2), (10.7), and (10.16).

Table 10.3 shows the productivity of several proposed products relative to that of an existing double level metal gate array chip of the same size and with the same ground rules, and assumed to have yield of 25%.

Table 10.3
Relative Productivity for Different Designs

P_{dmw}	$d = 1$	$d = 2$	$d = 3$
$m = 2$	1	3.8	0.69
$m = 3$	0.79	0.71	0.4
$m = 4$	0.5	0.4	0.15

Although standard cell and microprocessor design systems give chips with more circuits per unit area than can be obtained with a gate array design, they also have lower yield, and thus lower relative productivity.

It is sometimes asked, "What size chip will optimize productivity?" Unfortunately, this is not a well-posed question, since the productivity depends not only on chip size, metal levels, ground rules, and design system, which are fixed once production starts, but also on the defect density, which is a time-dependent term, in that efforts are constantly being made to lower it by the use of better tools and fabrication techniques. As a result, a chip size that optimizes productivity for one combination of design parameters in one time period, may be far from optimum in another. In fact, a detailed analysis of this question [2] shows that, during a given time period, it is possible for there to be more than one combination of design parameters and chip size that maximizes productivity. However, the expressions given in this section make it possible to make informed decisions on which of the competing products it will be advantageous to fabricate in a particular time period.

10.7 AVAILABILITY

In addition to productivity, there is one more crucial quantity, and that is availability or timeliness. A product that is not available in time cannot be successful. A product that does not achieve adequate yield will also not be successful. Therefore, it is important to be able to assess how long it will take from start of normal production for a proposed product to achieve a specified yield. This is determined by equating the specified yield Y_{spec} to the yield predicted by Equation (10.16) q quarters after the proposed product has started production, and then solving for q. So, if

$$Y_{spec} = Y_{12w_0}(s_0)^{\alpha \Gamma_{dmw} 2^{-q/\tau}} \tag{10.18}$$

then

$$q = \frac{\tau}{\ln 2} \ln[\alpha \Gamma_{dmw} \bar{Y}] \tag{10.19a}$$

where

$$\bar{Y} = \frac{\ln Y_{12w_0}(s_0)}{\ln Y_{spec}} \tag{10.19b}$$

from which, using the terms that comprise the design system and area scale factors, it is seen that

$$q = \frac{\tau}{\ln 2} \left[2(1 - b) \ln \frac{s}{s_0} + \ln f_d + \frac{(m - 2) \ln 2}{2} + 2 \ln \frac{w_0}{w} + \ln \bar{Y} \right] \tag{10.20}$$

If it is assumed that the learning rate is the same for the different proposed products, then by inserting into Equation (10.20) the appropriate values of the indicated parameters, it is possible to calculate the time that is expected to be needed to achieve the specified yield. An expression such as this makes it possible to see the effect each term has on the availability of a proposed product, and provides guidance on actions that need to be taken to reduce this time.

10.8 CONCLUSION

Decisions on product strategy are influenced considerably by the expected cost of a product and the time that will be needed for it to be available. A major contributor to the cost is the actual fabrication of the product. Depending on the design, the absolute cost of different products may be different, so meaningful comparisons need to be made on a unit cost basis (i.e., cost per packaged chip or cost per available circuit). This cost is determined by the fraction of potentially available chips that are indeed available, by the number of circuits on a chip (i.e., by the yield and the productivity) and by the availability. Accurate prediction of all of these are central to a successful business.

10.9 REFERENCES

[1]   A.V. Ferris-Prabhu, "Design Considerations for Yield," *Proceedings of RELECTRONIC '91*, 8th Symposium on Reliability in Electronics, Budapest, Hungary, Vol. I, August 26–30, 1991, pp. 868–880.

[2]   A.V. Ferris-Prabhu, "Parameters for Optimization of Productivity at Wafer Level," *IEEE Transactions on Electron Devices*, Vol. 39, No. 4, April 1992, pp. 952–958.

[3]   A.V. Ferris-Prabhu, "An Algebraic Expression to Count the Number of Chips on a Wafer," *IEEE Circuits and Devices Magazine*, Vol. 5, No. 1, January 1989, pp. 37–39.

About the Author

Albert V. Ferris-Prabhu has been a member of the technical staff of the IBM General Technology Division, Essex Junction, since 1968, and an adjunct Professor at the University of Vermont since 1978. He earned his MS in engineering from Princeton University and his PhD in solid state physics from the Catholic University of America. He completed his post doctoral work at MIT and the NASA Goddard Space Flight Center. His research in the areas of applied mathematics, magnetism, device physics, reliability, and semiconductor yield prediction has been reported in over 100 publications. He is a Fellow of the American Association for the Advancement of Science and a senior member of the IEEE.

About the Author

Index